CORPUS

CORPUS

AN INTERDISCIPLINARY READER ON BODIES AND KNOWLEDGE

EDITED BY
MONICA J. CASPER
AND
PAISLEY CURRAH

palgrave
macmillan

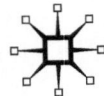

CORPUS

First published in 2011 by
PALGRAVE MACMILLAN®
in the United States—a division of St. Martin's Press LLC,
175 Fifth Avenue, New York, NY 10010.

Where this book is distributed in the UK, Europe and the rest of the world,
this is by Palgrave Macmillan, a division of Macmillan Publishers Limited,
registered in England, company number 785998, of Houndmills,
Basingstoke, Hampshire RG21 6XS.

Palgrave Macmillan is the global academic imprint of the above companies
and has companies and representatives throughout the world.

Palgrave® and Macmillan® are registered trademarks in the United States,
the United Kingdom, Europe and other countries.

ISBN: 978–0–230–11380–0

Library of Congress Cataloging-in-Publication Data

Corpus : an interdisciplinary reader on bodies and knowledge / edited
by Monica J. Casper and Paisley Currah.
      p. cm.
      ISBN 978–0–230–11380–0
      1. Human body—Social aspects. 2. Human body—Symbolic aspects.
  I. Casper, Monica J., 1966– II. Currah, Paisley, 1964– III. Title. IV. Series.

GN298.C68 2011
128′.6—dc22   2010043585

A catalogue record of the book is available from the British Library.

Design by Newgen Imaging Systems (P) Ltd., Chennai, India.

First edition: June 2011

10 9 8 7 6 5 4 3 2 1

CONTENTS

ACKNOWLEDGMENTS

WE WISH TO ACKNOWLEDGE the generous intellectual contributions of Jeffrey Bussolini, C. L. Cole, Lisa Jean Moore, Jackie Orr, Victoria Pitts-Taylor, and William Paul Simmons. Their suggestions, critical insights, and encouragement were instrumental in the transformation of our initial inquiry into a finished project. We also wish to thank our intrepid authors, who patiently reworked their excellent and diverse chapters to make this collection gel into a shared conversation rather than an assortment of disparate arguments. We are grateful to Brigitte Shull and Joanna Roberts at Palgrave Macmillan for their sound editorial guidance, and to artist Douglas Prince for use of his brilliant image on the book's cover.

Monica would like to thank Mason Olivia and Delaney Rose for their extraordinary patience while this book was being completed; there were a few too many occasions on which their mommy had to forego fun and silly activities for Serious Work. Fortunately, Bill Simmons was ever-present to hold down his end of the domestic teeter-totter while also engaging in his own vital intellectual pursuits; he is largely responsible for the relatively smooth functioning of our household, the endless cups of sweetened Awake tea that fueled my harried mornings, and the happy circumstance that neither daughters nor canines (nor I) went for too long without regular feedings. At ASU, Monica is blessed with an exceptional staff in the HArCS Division; a big shout out to Mary Bauer, Lucy Berchini, Tracy Encizo, Jackie Gately, Cathy Kerrey, Robert Kiec, Sarah O'Neal, and the student workers who keep the front desk and the "West Wing" humming. And as always, thanks to my family of origin for their humor and loving support, and to my intellectual compañeras Adele Clarke and Lisa Jean Moore for two decades of interlocution, friendship, and embodied empathy.

Paisley would like to thank the students, staff, and colleagues at Brooklyn College whose commitments to the work of higher education, and pursuit of its joys, make the college a wonderful place to teach and to write. He is especially grateful to Barbara Haugstatter in the Political Science Department. From tracking lost packages to getting a photocopier fixed, her labors did and do much to lessen his. The engagement and unbounded questioning of the students in his senior seminar on bodies and biopolitics turned that class into the most rewarding reading/writing group he has ever been in. He is indebted to PSC-CUNY Research Award Program for supporting his work curating this collection. As always, he is deeply grateful to Shannon Minter for his intellectual companionship over the years and to Kelly Currah for his unwavering enthusiasm. Finally, he is delighted to acknowledge the love and support of his family. Grace, Georgia, and Greta warp adult time in surprising and beautiful ways. And Lisa Jean's magical way of creating order out of chaos makes his work possible. She never ceases to amaze.

CONTRIBUTORS

Monica J. Casper is professor of women and gender studies and director of the Humanities, Arts and Cultural Studies Division at Arizona State University's New College. She is author of *The Making of the Unborn Patient: A Social Anatomy of Fetal Surgery*, editor of *Synthetic Planet: Chemical Politics and the Hazards of Modern Life*, and coauthor of *Missing Bodies: The Politics of Visibility*. She is currently researching the biopolitics of infant mortality and maternal-child health in the United States and globally.

Paisley Currah is professor of political science at Brooklyn College of the City University of New York. His current book project, *The United States of Sex: Regulating Transgender* (forthcoming, NYU Press), looks at contradictions in state definitions of sex. He is also a coeditor, with Shannon Minter and Richard Juang, of *Transgender Rights* (University of Minnesota Press 2006).

Darcy A. Freedman is an assistant professor and Centenary Faculty in Social Disparities in Built Community Environments at the University of South Carolina College of Social Work. Her research seeks to create healthy contexts and lives using community-engaged and participatory approaches. She has worked collaboratively with communities to develop farmers' markets, community gardens, and a grassroots policy coalition focused on food justice.

Jonathan Xavier Inda is associate professor of Latina/Latino Studies at the University of Illinois, Urbana-Champaign. His publications include *Targeting Immigrants: Government, Technology, and Ethics* (2006) and the edited volumes *Race, Identity, and Citizenship* (1999), *Anthropologies of Modernity: Foucault, Governmentality, and Life Politics* (2005), and *The Anthropology of Globalization*, 2nd Edition (2008).

Stephen Katz is professor of sociology at Trent University in Peterborough, Canada. He is author of *Disciplining Old Age* (1996), *Cultural Aging* (2005), and several book chapters and articles on critical gerontology and the aging body. Currently, he is working on the cultural aspects of aging memory and cognitive impairment.

Mary Kosut is assistant professor of Media, Society, and the Arts at Purchase College, State University of New York. Her books include *The Body Reader: Essential Social and Cultural Readings* coedited with Lisa Jean Moore (NYU Press) and *The Encyclopedia of Gender and Media* (Sage Press, forthcoming). Her research focuses on the body, visual art, consumption, and popular culture. She has published on tattoo art, body modification, and academic culture in *Deviant Behavior, Visual Sociology, The Journal of Popular Culture*, and *Cultural Studies-Critical Methodologies*.

Kathleen LeBesco is professor of communication arts and distinguished chair at Marymount Manhattan College. She is author of *Revolting Bodies? The Struggle to Redefine Fat Identity*, coauthor of the forthcoming *Culinary Capital*, and coeditor of *Bodies out of Bounds? Fatness and Transgression, Edible Ideologies: Representing Food and Meaning, The Drag King Anthology*, and special issues of *The Review of Education, Pedagogy and Cultural Studies* on the teacher's body, and *Women and Performance: A Journal of Feminist Theory* on excess.

Shoshana Magnet is an assistant professor in the Institute of Women's Studies/l'Institut d'études des femmes and the Department of Criminology at the University of Ottawa. She is completing a book titled *When Biometrics Fail: Culture, Technology and the Business of Identity* on how new identification technologies encode assumptions about race, gender, and sexuality. Her published work appears in journals such as *Feminist Media Studies, Qualitative Inquiry, New Media & Society*, and *Body & Society*.

Diana Mincyte is a fellow in the Program in Agrarian Studies at Yale University. Her research explores topics at the interface of poverty, consumption, biopolitics, and the environment, particularly in the contexts of Eastern Europe and the Global South. Her book project focuses on raw milk politics in postsocialist Europe to consider connections between subsistence practices and sustainability discourses.

Lisa Jean Moore is professor of sociology and gender studies at Purchase College, State University of New York. She is author of *Sperm Counts: Overcome by Man's Most Precious Fluid*, coauthor of *Missing Bodies: The Politics of Visibility* and *Gendered Bodies: Feminist Perspectives*, and coeditor of *The Body Reader: Essential Social and Cultural Readings*. Currently she is collaborating on a cultural interpretation of human-bee relationships.

Michelle Murphy's scholarship explores questions of technoscience, life, and economy in the late twentieth century. She is associate professor of history and women and gender studies at the University of Toronto. Her books include *Sick Building Syndrome and the Politics of Uncertainty: Environmental Politics, Technoscience, and Women Workers* (Duke University Press, 2006) and the forthcoming *Seizing the Means of Reproduction: Technoscience, Feminist Health, and Biopolitics in the Contradictions of American Empire.*

Dylan Rodríguez is professor and chair of the Department of Ethnic Studies at University of California, Riverside. He is author of the books *Suspended Apocalypse: White Supremacy, Genocide, and the Filipino Condition* and *Forced Passages: Imprisoned Radical Intellectuals and the U.S. Prison Regime.* He is a founding member of Critical Resistance: Beyond the Prison-Industrial Complex.

Lara Rodriguez is a PhD candidate in the English program at The Graduate Center, City University of New York. She is currently investigating transdisciplinarity through and across theories of desire, eros, and electricity.

George Sanders is assistant professor of sociology at Oakland University. His research focuses on the production of religious spectacles as well as the more general ways contemporary consumerism connects with traditional, sacred rituals.

Maurice E. Stevens is associate professor in the Department of Comparative Studies at Ohio State University. His research interests include critical trauma studies, theories of embodiment, performance, critical psychoanalytic studies, ethnic and gender studies, and American studies. His first book was *Troubling Beginnings: Trans(per)forming African-American History and Identity,* and his forthcoming book project is *From the Past Imperfect: Towards a Critical Trauma Theory.*

CHAPTER 1

BRINGING FORTH
THE BODY

AN INTRODUCTION

PAISLEY CURRAH AND MONICA J. CASPER

WE ALL FALL DOWN...

It is 2002. The US Central Intelligence Agency has captured Abu Zubaydah, believed to be a high-ranking member of *al Qaeda* involved in planning the attacks of September 11, 2001. The CIA wants to know if its plan for interrogating Zubaydah—involving a combination of "walling," facial holds and slaps, confining him to a box, putting insects in the box with him, sleep deprivation, "wall-standing" (standing a few feet from a wall with one's arms stretched and one's fingers on the wall), using various "stress positions," and "waterboarding"— would violate a federal law prohibiting torture. Two *doctors of jurisprudence* in the Attorney General's Office of Legal Counsel, Jay Bybee and John Yoo, draft a memo explaining precisely why the CIA's planned "enhanced interrogation techniques" do not constitute torture.[1] Much has already been written about this "torture memo" and others released under Freedom of Information Act (FOIA) requests.[2] But one aspect of Bybee and Yoo's argument is especially salient here.

The US statute that makes torture a crime defines it as "an act committed by a person acting under the color of law specifically intended to inflict severe physical or mental pain or suffering…upon another person within his custody or physical control."[3] The authors interpret this to mean that, to be guilty of torture, one has to actually *cause* severe pain or suffering and to *intend* to do so. For "intent" to be present, the interrogator would have to possess *knowledge* of the possible severe effects of these techniques on the victim's body or mind. According to Bybee and Yoo, there was no such intent: "We have no information from the medical experts you have consulted that the limited duration for which the individual is kept in the boxes causes any substantial pain"; "Based on your research into the use of these methods at the SERE [Survive, Evade, Resist, Escape] school and consultation with others in the field of psychology and interrogation, you do not anticipate that any prolonged mental harm would result from the use of the waterboard." Bybee and Yoo make the case repeatedly that "reliance on the advice of experts"—who allegedly have no information that severe pain will result—establishes a "good faith belief" that the interrogator did not intend to violate the law, and so these techniques, used singly or in combination, do not constitute torture.

As a result of the official memo, Zubaydah and others were subject not to "torture," but to a variety of ostensibly benign "interrogation techniques." We learn later from one of the interrogators that "no actionable intelligence [was] gained from using [these] enhanced interrogation techniques."[4] Of course, invoking the argument—in fact, the widespread consensus—that torture does not work plays into an instrumental logic suggesting that there are some cases in which torture is justified. This dissolves the foundations of the principal, enshrined in the UN Convention against Torture, which says states should not torture. Indeed, in another secret memo, Justice Department attorneys argued that, even if the interrogations *did* constitute torture under US law, the president's power as commander in chief to protect "the security of the nation" trumps laws protecting the safety and integrity of the individual body.[5] So we now operate under an apparatus of "national security" in which presidents and their agents behave like sovereigns of yore, screaming "Off with their heads!"— and actually cleaving bodies.

What are we to make of this horribly banal moment, in which scholarly institutions inaugurated and sheltered over the centuries to extend the boundaries of what can be thought and known, to advance human progress, were able to bring forth a body of knowledge about the body that is used to justify its torment? After centuries of study and accumulation of massive amounts of data about the effects of various harms and deprivations to human (and nonhuman) bodies, how is it that medical and legal "experts" credentialed through the self-governing apparatuses of professionalism and state licensing boards can state they have "no information" that interrogation methods will cause severe physical or mental pain? What enables us to bury evisceration inside of abstraction, conceal annihilation inside of so-called rational action, and rupture bodies and lives in the service of (re)building nations and empires? Surely these post-9/11 legal machinations comprise the sordid underbelly of the Enlightenment's bright shiny carapace?

To begin answering these questions, let us travel back in time to the origins of English common law. The phrase habeas corpus, whose foundations were laid in the 1215 Magna Carta, means "you (shall) have the body." Yet the phrase has typically been transposed into the injunction to "bring forth the body." The writ's "most great and efficacious" version, William Blackstone explains, is the writ of habeas corpus *ad subjiciendum* "directed to the person detaining another, and commanding him to produce the body of the prisoner with the day and cause of his caption and detention."[6] Being able to order authorities to bring a body before the court to justify "its" confinement as lawful has been heralded as "a great security," a confined person's last chance to prevent a miscarriage of justice in legal systems descending from English common law.[7] No secrecy, no languishing in prison hellholes without chance of ever presenting one's case to an impartial court, and no "extralegal" state-sponsored kidnappings, torture, or execution.

The era that witnessed the beginnings of the idea of habeas corpus also marked the establishment of the great medieval European universities.[8] In these organized sites of knowledge production, the "corpus" to be produced was not a human body, but rather a body of knowledge—profoundly disembodied knowledge, as it later turned out.[9] Universities were literally corporations, "associations of students and teachers with collective legal rights usually guaranteed by charters issued by princes, prelates, or the towns in which they were located."[10] Historian Marcia Colish notes that universities and their subcorporations, the individual faculties (schools, or what we might now call departments), mirrored their counterparts in the marketplace, the craft guilds:

> Like guilds, universities were self-regulating. They determined their own rules for membership and for marking the levels of competence required for advancement to higher grades of activity. The main reason why universities sought, and gained, legal independence was to ensure that promotion was based on academic competence, as verified by masters with the requisite expertise...The resultant liberties gave medieval universities far more academic freedom.... [and] enabled faculties, whoever endowed them, to set their own standards and to modify and update curricula in light of new intellectual developments and materials.[11]

Ivied walls of universities provided shelter for the production of knowledge against the *doxa* of princes, legislatures, popes, and publics. No longer constrained by common sense limits on the thinkable, newly able to question that which "goes without saying,"[12] universities flourished, reproduced, and engendered systematic bodies of knowledge in dozens of disciplines, many hundreds of subfields, and thousands of highly specialized areas. These bodies of knowledge became institutionalized, codified, and literally inscribed into books and journals, divisions and schools, professional associations, elaborate credentialing systems, and intellectual legacies.

Incorporation of the university and its disciplinary subdivisions was to be a bulwark against interference in the pursuit of knowledge. But the production and organization of knowledge within academic architectures has created its own myopias, for knowledge is relentlessly political. As a knowledge project, Foucault points out, a discipline is "essentially centripetal." That is, it "isolates a space...determines a segment...concentrates, focuses, and encloses."[13] And it is quite often the organic body that becomes the object to be contained, manipulated, and "rendered both useful and docile."[14] Each discipline projects its own idealization of the human or animal body, and this image reflects back on the discipline itself, defining the parameters, the shape, the density, of the project—its "proper" object of analysis, the questions that must be asked, and the questions that cannot. The centripetal forces—cultural, institutional, economic, political, and symbolic—that structure disciplinarity produce a coherent understanding of the body that differs for each field; indeed, what is an axiomatic truth for one discipline might be absurd or unintelligible for another.

What is gained and what is lost when the commons of the body is enclosed? Disciplinary techniques, such as those legal renderings described earlier, squish and mold the biological material of a living being into a neatly bounded container—the individual—that can show up for work, take a vacation, resist through the ballot box, keep a satisfactory level of personal hygiene, do whatever is proper given their station, and be subject to "enhanced interrogation techniques" and other social controls. These same disciplinary techniques turn human bodies and tissues (living and dead) into aggregate measures and professional fields, such as demography and epidemiology.[15] Imperatives to create a legitimate corpus of knowledge have ossified into intricate discursive structures with the potential to produce contorted depictions of the body that our early modern forebears, despite the cruelty wrought by their own nation-states, would have trouble parsing through.

ORIGIN STORIES

This collaborative project originated both from a certain intellectual dissatisfaction—born of our sustained, sometimes stalled labors to engage in interdisciplinary scholarship—and from an optimistic leap of the imagination. We were motivated by the hope that scholars within and across radically different disciplines might be able to talk to each other. And even more that these scholars might learn to create new knowledge together. Of course, "breaking down silos" is today's buzzword phrase, and many scholars frequently engage in interdisciplinary work. Yet we often do so from within the familiar (read: dysfunctional and restrictive) spaces of our "home" disciplines. We may be more comfortable operating within the parameters of our taxonomic divisions, social scientists consorting with social scientists, humanists with humanists, life scientists with life scientists, and so on. Further, while interdisciplinarity may be on the rise, it remains deeply troublesome—both to the academy, which still does not quite know what to do with it, and to individual scholars who want to branch out but are uncertain or simply cannot overcome their disciplining. When we do transverse established intellectual borders, it is often for an idea here, a method there, like Lévi-Strauss's *bricoleurs*, rather than full immersion and expertise in another scholarly field.

Many scholars have on their bookshelves excellent volumes that bring together different disciplines around a particular subject or topic: culture, sexuality, the body, the environment, science and technology studies, and disability, to name just a few. What typically allows these volumes to hang together is the topical focus; new theories and methods are advanced that add to our knowledge about the subject in question. However, while we may learn fresh ways of examining important topics, such as sexuality or disability or pollution, we often do not learn as much as we would like about the author's field or discipline of origin and, more importantly for our purposes here, his or her knowledge-production practices. There is limited self-reflexive dialogue in many edited volumes about *how* studying a central theme, whether sexuality or the body, contributes to new configurations of disciplinary or interdisciplinary knowledge. This may be too broad an epistemological claim given the surge of interdisciplinary scholarship, but we suggest the time may be ripe for a volume focused explicitly and expansively on the conceptual and methodological practices of interdisciplines. How, in other words, do we actually *do* interdisciplinarity, with and for whom, about what, and with what consequences?

We demonstrate in the following pages that the edifice of any particular knowledge project is built around the specter of the body it calls forth (or erases); that vision in turn sets its ontological, epistemological, and methodological limits. In preparing this volume, we were determined to include diverse interdisciplinary fields and intellectual spaces—some widely recognized, some freshly emergent, many of which had not yet intersected. These ranged from death studies to fat studies to feminist science and technology studies to trauma studies, with myriad in between, as we deliberately wanted to put into conversation fields that *infrequently* engage each other. Interested in the project of interdisciplinarity, we charged our authors with addressing their respective fields—its origins, key concepts, problematics, ideas, methods, and controversies—through the heuristic of the body and embodiment. In other words, we said, tell us about your field, but do so through a story or stories about the body. *Corpus* thus channels a specific conceptual thread, namely how each scholarly area discussed here has been shaped by intellectual emphasis on the body and is equipped (or not) to address embodiment.

It does not take long for novel interdisciplines to erect their own edifices, to encase the body within new methodological limits and normalizing gazes, and to set up walls against the incursions of even more nascent interdisciplinary *arrivistes*. Yet, at least initially, in the small and perhaps ephemeral zones between the academy's disciplinary architecture, in those fortuitous moments of vertigo before certitude is restored, interdisciplinary work can conjure previously unthought-of yet incredibly exciting and productive combinations. In these formations, the body is increasingly dispersed away from what was thought to be its center, the individual human. And so the study of the body calls for critical projects that are not bound to particular locations—be they anatomical, geopolitical, epistemological, or ontological—but rather can be pursued across multiple diasporas. In short, bodies do matter but have sometimes disappeared in tensions invoked by epistemological and ontological battles over the primacy of the social or the inevitability of the biological, and in massive amounts of discourse and data that divorce us, sometimes quite messily, from the body/our bodies.[16]

Corpus is no exception: the body itself is not actually present in these pages although we encouraged our authors to always keep the body in view, to *represent* its materiality in fresh and exciting ways. Textual investigations do not just limit, but establish a particular grid that enables certain kinds of questions and makes others impossible. Other sorts of scholarly and creative interventions—performance pieces, film, digital video—would enable different questions and impose different limits, yet even digital media can dissolve the body in its endless reproductions. But these are not the media in which most of us work. In limning the body, we still traffic in words and symbols, which are always already not the lived body. We did not create a multimedia installation here but a book; this corpus was produced through email conversations, phone calls, tracking changes in documents, and disembodied presences scattered and almost instantaneously reassembled into zeroes and ones across thousands of miles. This collection has not "found" the body or wholly delineated its new parameters, but shown how attempts to fix it in space and time, to circumscribe it as only an effect of language, foreclose critical analysis.

We have organized the book's diverse, interrelated chapters by merging a kind of "organic" approach with a generational approach, working across the "naturalized" life-course from conception to birth to death, and through the body, its boundaries and processes, from inside to out. At the same time, we grouped chapters together based on shared themes, moving from reproduction, through race and biology, to food and fatness, to disabled, sexual, and laboring bodies, to new media and digital bodies, to trauma, aging, and death. Collectively, the chapters herein address head-on relationships between theory and phenomena, objects and subjects, and the "natural" and the "social." They attempt to account for bodies in space and time, scrutinizing data, practices, particularities, and material specificities. Singly and together, the chapters show that bodies produced through power relations are not necessarily unitary targets, bound by one skin, having one heart, with one individual installed as tenant. Instead, the twenty-first century body is sliced up in new ways, rearranged and redistributed across much broader temporal and spatial configurations. In expanding research on the body to take into account phenomena at diminishingly micro and exceedingly macro levels—from genomes to nanos to organs to groups to populations—and within and across species, our authors do something analogous to cutting off the king's head. That is, they set aside processes of subjectification (at least analytically), and instead see the body as a process, a target, a commodity, an avatar, an audible piece of inventory circulating through public and private spaces—but not as always or even centrally a tracing pattern for individuation.

OVERVIEW OF THE CHAPTERS

We begin, as does most human existence and a great deal of cultural politics, with reproductive bodies. Historian Michelle Murphy shows that the reproductive body is neither macro nor individual, but rather extensive across and constitutive of diverse domains. A distributed ontology of reproduction based on what Catherine Waldby and Robert Mitchell called tissue economies serves to redefine human reproduction while at the same time fragmenting reproducing bodies, dispersing their reproductive parts as risks or commodities across social networks.[17] Engaged with feminist technoscience scholarship, Murphy examines the contested terrain of biological human reproduction. She coins the term "distributed reproduction" to reveal how global and local practices generated through transnational networks have fostered "designations of lives more and less productive, lives unworthy of birth." Tracing the evolution of reproductive epistemologies from natural sciences and political economy since the eighteenth century, she examines how formations of reproduction and formations of capital were co-constitutive with significant consequences for the present. Specifically, Murphy investigates two contemporary instances of distributed reproduction in action.

In the first case, "The Girl Effect," a development campaign run by the Nike Foundation and the Novo Foundation proposes investing in girls as a means of rescuing nations from devastating poverty. Her analysis reveals how the campaign of fertility reduction directed at these specific girls has become a proxy for correlations with economic productivity related to nation-building. The second case, the production of bisphenol A (BPA), a plastics additive associated with toxicity, specifically estrogenicity, showcases the Canadian "chemical alley" of Sarnia. Murphy locates BPA in the ghostly bodies of lost generations of boys, whose birth rates have been dramatically reduced through chronic chemical exposure of their mothers and grandmothers. Murphy's comparative analysis of distributed reproduction demonstrates that "living being" is increasingly conceived as alterable and materially transformable, opening new possibilities for "a malleable ontology of life."

Ethnic studies scholar Dylan Rodríguez's contribution, "Multiculturalist White Supremacy and the Substructure of the Body," targets the breach between biological and social narratives from an entirely different perspective, yet also addresses issues of chronicity and malleability. In his account, race matters but now it is untethered from particular biological living beings. Against the backdrop of celebrations attending Barack Obama's presidential rise and the allegedly "postracist" inauguration, Rodríguez suggests that "bodily desegregation" is not the end of white supremacy, but rather its crowning achievement. He interrogates a central contradiction: How is white supremacy "fully engaged in the organization of systemic violence and the social subjection of white civil society's historical racial antagonists, *even and especially* as its institutional forms—including its political and intellectual leadership—display a (relative) 'diversity' of nonwhite bodies?"

Examining Obama's rhetoric alongside the Los Angeles Police Department's embrace of "phenotypic diversity," Rodríguez suggests that "erosion of the 'white body' as the singular prototype of racist and white supremacist agency" has not in fact displaced white supremacy. He rebuts the notion that racism is merely epiphenomenal to other, more important historical formations and structures. Instead, he argues, white supremacy sits alongside capitalism (and patriarchy) as its own historically dynamic logic of social organization. He shows the necessity of a new critical approach to race and white supremacy—one that moves from phenotype to structural analysis, and that frames the body in novel ways. With the "political *and bodily* desegregation of white supremacy," we see the emancipation of the racialized body from the political projects that fostered and contained it, including critical race studies. White supremacy thus becomes even more purely installed—no longer need it rely on the blunt instrument of Bull Connor.

Focused, like Rodríguez, on the potent socioorganic nexus of race, anthropologist Jonathan Xavier Inda, in "Materializing Hope: Racial

Pharmaceuticals, Suffering Bodies, and Biological Citizenship" shows that while race is not inherently genetic, the biological effects of racism cannot be ameliorated solely through actions in the social realm. Inda's analysis centers on debates around the drug BiDil, which researchers found to have disproportionately positive effects as heart failure therapy for African Americans and marketers deployed via racially targeted advertising. Because, Inda argues, social inequality is materialized *biologically* among African American communities—in higher rates of heart failure, in longstanding "biomedical neglect" in drug effectiveness, and in access to health care—the "vital needs and suffering of the black body" must be part of the solution. He uses Foucault's concept of biopower to disentangle the racist practice of geneticizing African Americans (the most recent progeny, he argues, of the racial sciences) from the "biological citizenship project" of lessening health disparities and achieving "vital rights." Here, bodies are productive opportunities rather than foundational grounds for political action.

Moving from explicitly racialized bodies to consuming bodies (which may also be racialized), in "Embodying Food Studies: Unpacking the Ways We Become What We Eat," public health and community studies scholar Darcy A. Freedman interrogates the ostensibly simple question of what people should "have for dinner." Aiming for a more contextual, complex, and embodied study of food and food practices, Freedman presents three engaging case studies—a middle school cafeteria in Louisiana, a farmer's market in Nashville, and an episode of television's *The Wire*—to examine stratified experiences of consumption. She shows that food in all its facets, from its origins to its visual display in commercial markets to its representations in knowledge structures to its ultimate ingestion, is profoundly political. Her account works to carve into the supposed divide between biological material and social processes, identifying these as always and already interconnected, and indeed revealed through people's bodies and their (attempted) achievement of basic human survival. Her call for an embodied food politics challenges entrenched dualities and argues for a more complicated, multidimensional understanding of the relationship between what we eat, what we know, and how we live (and die).

The relation between subjectivity and embodiment is critical to scholarship in which "deviant bodies" are objects of analysis.[18] Thus, it is only partly coincidental that food studies and fat studies are contiguous in *Corpus*, for both reveal body politics in strangely familiar ways. Much work in emergent Fat Studies strives to understand how cultural imperatives to have/be a thin body are increasingly biomedicalized. In "Epistemologies of Fatness: The Political Contours of Embodiment in Fat Studies," communications scholar Kathleen LeBesco charts how Fat Studies focuses on the slippage between cultural denigration of fat people and celebration of thinness, on the one hand, and association of obesity with ill health (an association not uniformly borne out by clinical data) on the other. For LeBesco, "Fat Studies involves not simply taking fat and fat people as an object of study, but instead seeing the fat body as characterized by subjectivity, which is itself a political project." In reclaiming "compromised subjectivities," LeBesco argues, Fat Studies has much in common with women's and gender studies, queer studies, and disability studies. These approaches center the study of bodies "received as transgressive, problematic, or dangerous" but invert the usual framing, focusing on the political, social, and cultural processes that thrust those bodies outside the body politic. For Fat Studies, however, this approach can too easily morph into environmental explanations in which the fat body is problematized from the perspectives of public health and food studies, and thus the fat subject is decentered once again. A cultural script that defines the fat body as the culpable result of bad human choices means that Fat Studies must navigate between the rock of "obesogenic environment" and the hard place of individual responsibility, between population-level biopolitical interventions and individual-level disciplinary "self-government."

Turning to sexualized bodies, sociologist Lisa Jean Moore and graduate student Lara Rodriguez, in "Identities without Bodies: The *New* Sexuality Studies," examine how (and why) the body has disappeared from sexuality studies. Beginning with sexuality studies in its earthier Kinsey-ish formations organized around the question, What do people *do, or want to do,* with their bodies?, they chart how, through the decades, these questions have been displaced by a focus on identity, on *who* is doing what to *whom.* For Moore and Rodriguez, "Something rather dire seems to have happened to the *fleshiness* of sexuality studies." Offering a content analysis of sexuality studies journals, they argue that "embodiment and bodies appear to have receded behind identity and power as core variables of interpretation and analysis." In the organization of new sexual knowledge formations, doing has slipped (back) into being, and identity has trumped practice. These shifts, they suggest, reflect a "cultural consumer practice that demands identifications in/by pre-determined groups before we even know our anatomies, pleasures, turn-ons, and turn-offs." With few exceptions, the field is now dominated by research on development of identities and subcultures rather than work that centers sex itself and the "verdant, carnal, sensate, drippy, leaky and engaged organism." For Moore and Rodriguez, robust descriptions of the body and its practices have been supplanted by a cultural discourse of signs and signifiers. One effect of this, they suggest, is a "marked neglect of women's sexuality and the embodied issues regarding female-bodied persons." While not essentializing "feminine" sexuality—however this may be defined—they nonetheless point to an elision of women's material sexuality in all its diverse configurations within identify-focused studies.

Of course, bodies do many things besides take pills, eat, and have sex. Environmental sociologist Diana Mincyte offers an insightful account of laboring bodies in "'The Bugs of the Earth': Reflections on Nature, Power, and the Laboring Body." Here, she draws on the example of Lithuanian subsidiary farming under Soviet rule to explore and trouble relations between natural and social apparatuses. In her framework, humans and nonhuman landscapes such as farms are both objects and subjects caught up in the historical, world-changing events of Soviet empire building; and both farmers and environments talk back—or are compelled to talk back through various techniques—to the state. As self-described "bugs of the earth," subsidiary farmers transformed the Lithuanian landscape through their everyday labor; Mincyte suggests that the farmers' embodied work, indeed their very bodies, structured their relations to the state. Laboring bodies served as links to state power and helped to solidify Soviet rule in and across twentieth-century Lithuania while also providing the material grounds and lived experiences for resistance to state authority. Mincyte revisibilizes aggregated "peasant populations" that are often erased in East European and environmental studies. In her work, bodies are central both to formations of political identity and to statecraft. And viewed through the lens of laboring bodies, the events of Soviet expansion look categorically different.

Connections between individuals and their bodies—as expressed, for example, through racialization, sexuality, and labor—are displaced (almost) entirely with Shoshana Magnet's analysis of subcutaneous Radio Frequency Identification (RFID) chips. Because surveillance had for so long been based on technologies that see the body when it enters a particular place (e.g., the Panopticon in the prison, the camera in the lobby, and the x-ray in the airport baggage check), surveillance studies has tended to rely heavily on visual metaphors and their fundamentally static economies of recognition. In "The Audible Body: RFIDs, Surveillance, and Bodily Scrutiny," communication scholar and art historian Magnet shows how, with the insertion of a chip, new surveillance technologies move from a machine fixed in one place, like a camera positioned on a wall, to the moving body itself. As a signal passing through space, its emissions circulating as pieces of data in networks of inventory, its discrete dative elements segmented and reassembled into other sorts of knowledges, the "audible body," Magnet argues, broadcasts information that would be invisible to the camera. With audible bodies, a different register of mobile knowledge supersedes the Enlightenment dream of visual transparency. It is no longer the self that speaks, that reveals its knowledge about the body, but rather the chip. And this material shift means the truths revealed by bodies now are of a different sort, capturing fluid assemblages of commodities and consumers, of inventories and workers, of populations and networks. As such, new ways of accounting for (and counting) bodies provoke new epistemological frames for knowing them.

Media scholar Mary Kosut takes as her research object spectral presences, or embodiment not limited by/to corporeality. In "Virtual Body Modification: Embodiment, Identity, and Nonconforming Avatars," Kosut examines the phenomenology of embodiment in the virtual social world "Second Life" (SL), centering her analysis on the "bleed between virtual and material embodiment." For Kosut, the finding that some users feel like their avatars—the virtual representations of themselves constructed in/for a digital world—suggests that ontological presence itself can be partially disarticulated from an embodied self and rearticulated in a digitized self. In SL, users feel and sense their incarnated avatars, making the body "present" even though it is absent, or external to the digital frame. Certainly, many users construct virtual selves that resemble their RL (real life) physical selves; others create selves that mimic the exaggerated ideals of perfection against which we are measured in RL. Kosut pushes the analysis one step further: What are the implications, she asks, of users who inhabit SL selves constructed *against* the imperatives and norms of RL? "The continuum of non-conforming avatars, whether it is a woman who rejects the Barbie ideal, an able-bodied person who embodies a disability, or vice-versa" show how people "use SL as a platform of resistance," a "space of transgression and interplay between ideal and real world norms." Such digital practices, like Magnet's RFID chips, extricate identities from bodies; but even more, they provide identities with new, virtual bodies to inhabit, scrambling our understanding of who somebody "really" is.

As we launched our project with the beginnings of human life, so we complete our collection of chapters with endemic human experiences of trauma, aging, and death. In "Trauma's Essential Bodies," comparative studies scholar Maurice E. Stevens offers a critical account of the emergence of trauma studies. He shows that theories of the body—including embodied notions of suffering, breach, and lived experience—have been essential to developing the field. Trauma studies as a relatively new interdiscipline registers traces of rupture in significant visible injuries and signs of damage. In such framings, trauma's power is "demonic," capable of "unmaking worlds" as Elaine Scarry wrote two decades ago.[19] Traumatic experiences are object and event at the same time, often diagnosed *through the body*. Trauma emerges as an apparatus, an assemblage of histories, bodies, identities, powers, and troubles, fostering specific forms of social life and meaning. Stevens lodges the history of trauma and its analytical projects first in railway accidents and then wartime wounding, each with its own "signature injuries." Such injuries, particularly Posttraumatic Stress Disorder (PTSD), instantiate the notion of bodies as texts through which harm is read. Yet cultural readings have created "fixed logics of the body"; the trace written on the body as harm becomes the guarantor that an event took place. But where and how do various traumas—specific, diverse, situated—live *outside* these fixed, taken-for-granted logics? Is the etiologic and epistemological category of trauma capacious enough to describe a range of bodily and psychic ruptures? Stevens suggests that, contrary to psychoanalytic presumptions, trauma does not merely describe embodied experiences, but rather creates them. He insists that analyses of trauma be attentive to heterogeneity, and that trauma is foremost a cultural phenomenon created from multiple events and molded within particular historical, geopolitical, and racialized apparatuses and contexts.

In "Hold On! Falling, Embodiment, and the Materiality of Old Age," sociologist Stephen Katz explicitly excavates what is buried when biological and social narratives are assumed to be fully incommensurate. He traces this fissure in relation to an incompatibility he identifies between the fields of aging studies and body studies. Katz argues that the "materiality of embodied aging" gets lost within both overly constructivist and narrowly medicalizing approaches. In body studies, important work challenging biologistic narratives has had the unfortunate effect of often ignoring the physical materiality of bodies. Indeed, embodied experiences of aging are often absent from social-constructionist work on bodies. On the other side of disciplinary divides, gerontology and geriatrics are heavily focused on biology and only slightly more attuned to biographic dimensions and subjective experience. Katz grounds his intervention on an examination of falling and fall prevention programs for the elderly. He makes a strong case for showing that falls can and should be understood through knowledge-making practices organized around the "coalescence of physical and biographical aging." Like experiences of aging themselves, aging bodies live (and sometimes fall and die) at the nexus of what we call the social and the biological.

Finally, in "The Gimmick: Or, The Productive Labor of Nonliving Bodies," sociologist George Sanders exposes how dead bodies, traditionally considered sacrosanct in many cultures, are now retailed for amusement in the West. In the funerary gimmick industry, dead bodies are transformed into commodities: headstones, vases, jewelry, sculpture, china, fireworks, reefs, pencils, oil paints, and more. Some bodies become the arch-typical fetish object (jewels); others are visually consumed in spectacular display (fireworks); some products take the materiality of the body and turn it literally into a signifier (paintings); and still others become objects that have both use value and exchange value (e.g., china, precious stones). Even after their subjects have given up the ghost, so to speak, dead bodies in whole or in pieces are deployed by clever industries, made to labor even in death and celebrated as commodities and objects of affection. These developments, tied to shifts in capital, are profoundly changing American death practices. In Sanders's account, dead bodies can now acquire new identities (although not always of their own choosing), just as they can be reconfigured into objects for others' consumption.

AN EXPLORER'S GUIDE

In describing above the individual chapters in the order of presentation, we made links between them. With our editorial bird's-eye view, we also suggest other possible routes for readers to navigate through *Corpus*. For example, you may visit chapters thematically, by favorite scholar, by searching the index for the familiar or the intriguing, or even randomly. Of course, you will make your own interpretations, finding key linkages, helpful articulations, and even gaps and omissions. To help direct you, though, we have pulled together an explorer's guide to the book, framing a handful of important overlapping themes: posthuman dislocations; relations between identities and bodies; events, apparatuses, and technologies; and the either/or of materiality and textuality.

Posthuman dislocations. As we discussed earlier, the notion of embodiment in the social sciences and humanities has been foundational for critically grounding body studies. In this scholarship, the Cartesian notion of "the ghost in the machine" has been supplanted by critical approaches that center the phenomenological concept of embodiment. As anthropologist Thomas Csordas explains, "Embodiment is an existential condition in which the body is the subjective source or intersubjective ground of experience." Embodiment studies, he suggests, "are not 'about' the body *per se*" but about "culture and experience insofar as these can be understood from the standpoint of bodily being-in-the-world."[20] Much of the scholarship in *Corpus*, however, moves beyond these somewhat obligatory nods to embodiment, complicating the notion, propelling it forward, or even rejecting it as the first point of entry into studies of the body.

Nietzsche's axiom, central to phenomenology, that there is no doer behind the doing, that the doer is constructed in and through the doing, has been absolutely central to studies of embodiment. Here, the body is not imagined as a biological entity distinct from the installation of self through various practices, apparatuses, and processes of subjectification. Yet these approaches to embodiment remain tied to liberal notions of an integral (and integrated) self. In some ways, the doer, the individual, the subject who is her own agent, remains in a privileged position in relation to the body—if not the owner at least the tenant of the biological material imagined as encompassing the self. Certainly, the self is constituted through action, processes, and practices, as symbolic interactionists have long argued.[21] But what of the technologies, commodities, and distributions that act on and through biological material *without* hailing the individual?

Giorgio Agamben suggests in a recent essay that we (whomever this imagined "we" is) now inhabit a moment in which collective processes of desubjectification are no longer coupled with (re)subjectification.[22] In fact, the body may be interpellated by various schemes and arrangements that can circumvent the self entirely. Disentangling the study of the body from that of the self reveals new biopolitical (and what Achille Mbembe calls necropolitical[23]) processes that produce commodities, track inventories, collect and collate data, and distribute reproduction, life, trauma, and death among collectivities, social networks, and populations. These distributive processes are also deeply stratified and increasingly "popular." Placing bodies at the center of stories of knowledge formation thus reveals how the notion of bodies as whole organisms—and indeed, their very integrity—is destabilized. Murphy's approach to "distributed reproduction" and Rodríguez's use of the concept of "bodily desegregation" to track racism in our "postracial" moment are apt demonstrations of one of the collection's plot arcs—that of disentangling the self from the body, and the whole body from its parts. We also see posthuman dislocations exemplified in Kosut's nonconforming avatars, Sander's commodified funerary gimmick, and Magnet's audible body.

Identities and bodies. Bodies have routinely disappeared into the maw of identity politics (e.g., feminism, black nationalism, lesbian and gay pride). Various social movements—such as civil rights, disability rights, and the women's health movement—linked bodies to discrete categories and lived experiences of race, gender, sexuality, class, disability, citizenship status, and more. Yet relationships among bodies, justice, freedom, and identities are far more complex than most "identity politics" has grappled with. The ascension of identity politics generated huge categorical problems, as revealed, for example, by the title of the 1982 classic, *But Some of Us Are Brave: All the Women Are White, All the Blacks Are Men: Black Women's Studies.*[24]

In the twenty-first century, the easy politics of identity that animated new social movements of the last century has been undermined. A range of theoretical displacements and political projects has unveiled the mechanisms that fasten bodies to identities: for example, liberal (white) feminism by women of color and transnational feminisms, lesbian and gay politics by queer theory, unitary single-issue accounts of identity by work on intersectionality, affect, and assemblage. Heterogeneous embodied experiences of marginalization do not always smoothly translate into tidy politics and clear courses of social and political action. Identity as corporeally fixed, in which the body is *the* site of injustice, tends (ironically) to ignore the diversity of human bodies and of embodiment itself as mixed, fragmented, parsed, contingent, changeable, and stratified. Thus, an unreconstructed identity politics that posits a predictable (and often unitary) relationship between bodies, subjects, and political projects has fallen out of favor, particularly in transdisciplinary spaces.

But in many disciplines, still, bodies remain either invisible or snugly tethered to and constitutive of identities. Certainly, in much cutting-edge scholarship, the "universal" body has rightly been displaced and particular historical bodies installed in its stead. Yet for the stories we tell about them to be culturally intelligible, fundable, and marketable, they have to refer to something or some*one* we already know—the human, the mother, the disabled, the lesbian, the fetus, the aged, the veteran, the executive, and the Other. These marked bodies have been, for at least two decades and with varied consequences, at the forefront of scholarship—as justifications for intellectual engagement, as objects of critique, and as (contentious) political subjects. We strive to show, however, that the bodies within these mobile, useful "straw" subjects and their specificities of function and practice are all too often minimized or deliberately elided. Important questions about the ways that subjectivities are secured to bodies, and to knowledge, are disregarded. We see this theme in Moore and Rodriguez's account of the entrenchment of identity even in approaches that position themselves as anti-identitarian, such as queer studies. It is also present in Mincyte's work on "bugs of the earth." She situates bodies not only as "mediators of social and political relations"—the body's role in the language of identity politics—but also "as material agents plugged into ecological systems." LeBesco's account of epistemologies of fatness works this question from another angle: why is the subjectivity of those inhabiting fat bodies so quickly and so easily erased or denigrated? Why have fat identities been contained by the language of illness, unlike queer subjects who vaulted the barrier between pathology and politics decades ago?

Events, apparatuses, and technologies. Philosophers routinely draw epistemological distinctions between events and objects, or between existence and occurrence. Yet because bodies as (inhabited) objects move through time and space (although not in uniform ways), their kinetic journey can be, and is, recorded in terms of events. But what counts as an event? Is it what happens to bodies, or what bodies do, or both? Is a body an event? Do events produce bodies? If objects and events are intimately connected, what defines the nature and/or site of the connection? Might such connections, in fact, be actualized in and through bodies themselves? Merleau-Ponty argued in *The Phenomenology of Perception* that bodies are both object and subject; perceptions of self and the world flow through the body's encounters with other objects in time and space. How might we (continue to) theorize such dynamics, and what are the stakes in doing so? How might technologies and apparatuses affect already messy relations between events and objects, and with what consequences? Stevens's examination of the construction of trauma as a one-time event, rather than as a long-term structural effect, is one important example of this particular plot line; Katz's work on falling, the body, and aging is another.

Moving beyond the either/or of body as material *or* body as text. Ironically, even in work that purports to center the body, such as aging studies and disability studies, the body has been neglected and made invisible in many knowledge projects, as Monica Casper and Lisa Jean Moore argue in *Missing Bodies*.[25] It circulates as a framing device, a target, an analytical tool, and now even its own field, but the body *in itself* remains underexamined and inadequately theorized outside of a still select literature (e.g., sociology, anthropology, cultural studies, gender studies). Bodies are endlessly contained by critical analysis, positioned as inert elements acted upon by disciplining apparatuses, and made "real" only through discursive projects, a kind of intellectual Velveteen Rabbit vivified by social interaction. All too often, however, the insistent matter of *actual* bodies in these fields disappears in the interstices of biological and constructivist epistemological frameworks. That is, bodies are understood as *either* wholly cultural fabrications and thus objects of disciplinary power, *or* as deeply material collections of fluids, tissues, and symptoms amenable to biopolitical intervention.

One consequence of this approach is that particular aspects of bodies, read as "markers" in some scholarship, become irreducibly lodged in *either* biology *or* culture. Race is a particularly volatile site of this materialization, as Rodríguez, Inda, and Stevens demonstrate in this collection. As an historical formation with real material embodied effects, race oscillates between the so-called biological and the so-called social—revealing that bodies are socialized and the social embodied *at once and the same time*. Consider, for example, the traumatic "strange fruit" of lynching or high African American infant mortality rates historically and today; both these embodied social phenomena are products of hierarchical racial formations and practices by which some marked bodies have been made to suffer, to the marrow, for the collective benefit of others.[26]

Many of the contributors to *Corpus* ask (and answer) an important question: Do "biological realities" exist beyond the boundaries of critical inquiry? Rather than conceptually making the body and its materiality (or rather, its multiple materialities) *invisible*, our authors largely refuse to read the body only as an effect, as produced in and through discourse. The stubborn *thingness* of bodies, their unpredictability and messiness, are read against knowledge paradigms that see them only as products ("the aging body") or targets ("the sick body") or supposedly obdurate realities ("the dead body"). Even as theoretical projects of the last half-century exposed different sets of techniques through which bodies become imbued with meaning, at almost every theoretical juncture they have tended to reinstall the body as passive and inanimate. The body, in these frameworks, is static and/or stable—not at all becoming. The chapters here engage in the challenging work of keeping bodies visible; and even more, keeping them dynamic to and within stories about knowledge formation and social and cultural processes. In this vein, Inda distinguishes the old racial sciences' biologization of race from a biological citizenship project grounded on corporeality that would help African Americans achieve their "vital rights" of "life, health, and healing." Freedman's approach to the study of food politics, which focuses on eaters as "agents in the production of socialized bodies," links biology with/in the arrangements of social inequality. Finally, Katz's refusal to read biological and social narratives about the aging body as incommensurate provides important insights into "the ways in which biography, culture, politics, and biology are braided together."

At the outset of this project, we were fully cognizant of the exquisite *carefulness* required to navigate intellectual spaces in which bodies of life exist side-by-side with bodies of death, bodies of terror consort with bodies of hope, and the pleasures of food, health, and sexuality traffic alongside the dangers of racism, trauma, and neoliberalism. In *Corpus*, as in the so-called real world, we show that militarized bodies exist alongside transgender bodies, animal bodies are coupled with human bodies, hungry bodies eat and sexual bodies play, injured bodies and nonhuman avatars are revealed by machines, embryonic bodies develop and aging bodies fall (and perish), persistently racialized and gendered bodies are subject to surveillance and control, and nonhuman bodies benefit and suffer from both mundane and extraordinary interactions with humans. In short, bodies live and die; communities grow and mourn; nation states thrive and fail. And we, as impassioned scholars living in this postmillennial, hyperdigitalized, globally warmed, deeply stratified, and achingly beautiful world, want to find ways to tell these and other stories.

NOTES

1. US Department of Justice, Office of Legal Counsel, "Memorandum for John Rizzo, Acting General Counsel of the Central Intelligence Agency," August 1, 2002. See also US Department of Justice, Office of Legal Counsel, "Memorandum for Alberto R. Gonzales, Counsel to the President," August 1, 2002.
2. See, e.g., David Cole, "Introductory Commentary: Torture Law," in *The Torture Memos: Rationalizing the Unthinkable*, ed. David Cole (New York: New Press, 2009), 1–40; Jane Meyer, *The Inside Story of How the War on*

Terror Turned into a War on American Ideals (New York: Doubleday, 2008); *The Torture Debate in America*, Karen J. Greenberg, ed. (Cambridge: Cambridge University Press, 2005).

3. 18 U.S.C.A. § 2340 (1).
4. Ali Soufan, "My Tortured Decision," Op Ed, *New York Times*, April 22, 2009.
5. US Department of Justice, Office of Legal Counsel, "Memorandum for Alberto R. Gonzalez, Counsel to the President," August 1, 2002.
6. William Blackstone, *Commentaries on the Laws of England*, Vol. 3 [Facsimile] (Chicago: University of Chicago Press, 1979), 131.
7. Justice Story, *Life and Letters of Justice Story*, Vol. 1, ed. William C. Story (Boston: Charles C. Little and James C. Brown, 1851), 491.
8. Blackstone suggests that habeas corpus, a "doctrine co-eval with the first rudiments of the English constitution," predated the Norman Conquest and was then "established on the firmest basis by the provisions of *magna carta*" in 1215. Blackstone, *Commentaries*, 133. The great medieval universities were chartered during the twelfth century. Marcia L. Colish, *Medieval Foundations of the Western Intellectual Tradition* (New Haven, CT: Yale University Press, 1999), 267.
9. For a historical/philosophical discussion of the micro-relationships of embodied knowers, such as Newton, Darwin, and Lovelace, to their disembodied scientific knowledge, see Christopher Lawrence and Steven Shapin, *Science Incarnate: Historical Embodiments of Natural Knowledge* (Chicago: University of Chicago Press, 1998).
10. Colish, *Medieval Foundations*, 267.
11. Ibid., 267–268.
12. Pierre Bourdieu, *Outline of a Theory of Practice* (Cambridge: Cambridge University Press, 1977), 166.
13. Michel Foucault, *Security, Territory, Population: Lectures at the Collège de France 1977–1978*, translated by Graham Burchell (New York: Palgrave Macmillan, 2007), 45–46.
14. Michel Foucault, *"Society Must Be Defended": Lectures at the Collège de France 1975–76*, translated by David Macey (New York: Picador, 2003), 249.
15. Monica J. Casper and Lisa Jean Moore, *Missing Bodies: The Politics of Visibility* (New York: New York University Press, 2009).
16. Judith Butler, *Bodies that Matter* (New York: Routledge, 1993).
17. Catherine Waldby and Robert Mitchell, *Tissue Economies: Blood, Organs, and Cell Lines in Late Capitalism* (Durham, NC: Duke University Press, 2006).
18. Jennifer Terry and Jacqueline L. Urla, *Deviant Bodies: Critical Perspectives on Difference in Science and Popular Culture* (Bloomington: Indiana University Press, 1995).
19. Elaine Scarry, *The Body in Pain: The Making and Unmaking of the World* (Oxford: Oxford University Press, 1987).

20. Thomas Csordas, "Embodiment and Cultural Phenomenology," in *Perspectives on Embodiment*, ed. G. Weiss and H. Haber (New York: Routledge, 1999), 143.

21. George Herbert Mead, *Mind, Self, and Society*, ed. Charles W. Morris (Chicago: University of Chicago Press, 1934).

22. Giorgio Agamben, *What Is an Apparatus? And Other Essays*, trans. David Kishik and Stefan Pedatella (Stanford, CA: Stanford University Press, 2009), 22.

23. Achille Mbembe, "Necropolitics," trans. Libby Meintjes, *Public Culture* 15:1 (2003): 11–49.

24. Gloria T. Hull, Patricia Bell Scott, and Barbara Smith, eds., *But Some of Us Are Brave: All the Women Are White, All the Blacks Are Men: Black Women's Studies.* (New York: Feminist Press, CUNY, 1982.)

25. Monica J. Casper and Lisa Jean Moore, *Missing Bodies: The Politics of Visibility* (New York: New York University Press, 2009).

26. See also Ruth Wilson Gilmore's work on the prison industrial complex. Racism, in Gilmore's important new construction, "is the state-sanctioned or extra-legal production and exploitation of group-differentiated vulnerability to premature death." Gilmore, *The Golden Gulag: Prisons, Surplus, Crisis and Opposition in Globalizing California* (Berkeley: University of California Press, 2007), 28.

CHAPTER 2

DISTRIBUTED REPRODUCTION

MICHELLE MURPHY

> Why should our bodies end at the skin?
> —Donna Haraway (1991)

> Invest in a girl and she will do the rest.
> —slogan of The Girl Effect campaign, Nike Foundation (1992)

> From this point of view a given amount of health impairing pollution should be done in the country with the lowest cost, which will be the country with the lowest wages.
> —Lawrence Summers, leaked World Bank Memo (1991)

WHERE DOES BIOLOGICAL REPRODUCTION RESIDE? "In bodies," is a probable response, perhaps framed by a birth story populated with genitals, sperm, eggs, kinship, family, contraceptive ethics, health care, and heteronormative futures that promise alignment with a "good life" of house, job, and affective bonds. Or the answer "in bodies" might be accompanied by claims to human rights coupled with critiques of coercive racist states and grief-filled accounts of lives lost due to negligence, violence, or scarcity. Tectonic forces conspire to offer "bodies" as an obvious answer to questions of human reproduction, not least because life and death hang on this process. Laws that direct reproductive rights and responsibilities to individual women, the medicalization of birth in all its uneven guises, and the "global facts of life" of economic development projects all converge to outline an individual embodiment for reproduction, drawing a perimeter around reproduction that has tended to designate the body of the possessive individual as reproduction's rightful home and liberal humanism as reproduction's proper imaginary terrain.[1]

What kind of epistemic concern has reproduction become? Yes, human biological reproduction is an embodied concern, posited through the right to bodily integrity, envisioned and managed through technoscientific practices, and experienced phenomenologically as a fleshly capacity shaped jointly by larger structural conditions (such as racial, capital, and social formations) and smaller micrological entities (cells and ribbons of nucleic acids precipitated for us by technoscience). Reproduction does happen in bodies, and lives become precarious in birth and pregnancy. But does reproduction stop there?

In this chapter, I want to address the question of what counts as biological reproduction by tracking the dispersion of sexed living being into its infrastructural and political economic milieu. To do so, this chapter considers how to (re)route reproduction through the social science infrastructures that wind the globe collecting data about birth rates, literacy, infant mortality, and microloans. It also considers the itinerary of bisphenol A as it is extruded by a plastic manufacturing process, dispersed through air, wafting here more than there, or leaching out of commodities into fluids that humans digest, collecting in bodies, human and nonhuman, disrupting metabolic processes, altering living being as it develops in time. In what sense do these two descriptions count as reproduction?

CONCEIVING REPRODUCTION

Before considering these two instantiations of reproduction, I want to spend some time tracking the intellectual and political stakes of conceiving of reproduction. As the anthropologist Marilyn Strathern points out, many of the very English words associated with reproduction—such as "conception," "creation," and "generation"—are also associated with acts of knowledge production.

Twenty-five years ago, Donna Haraway asked, "Why should our bodies end at our skin?" offering the "material-semiotic figure" of the cyborg as an ontological politics for attending to the ways living being was already constituted via technoscience in 1985, near the end of the cold war, in an emerging "informatics of domination."[2] At that moment, feminist technoscience studies was resisting a politics that posited bodies as natural entities and instead insisted that any question of "nature" or "biology" at the end of the twentieth century was already conditioned by technoscience.

In a similar spirit, one might pose the question, what is the ontological politics of embodied reproduction? In the same way that contemporary questions concerning the status of "sex" and "race" can be historicized as contestations over the terms of their existence—as contests over what kind of phenomena "sex" and "race" are—I ask: If they are materialized as "biological" forms, then residing where, with what boundaries, made manifest by what scientific techniques? If historical forms, explained by what conditions, at what scale? If social praxis, constituted through what relations, charted by whom? In other words, this chapter conceives of reproductive politics—and the question of what is reproduction—as a struggle over ontology, tracking not only how reproduction is hegemonically materialized in and through bodies, but also how countermaps of reproduction's uneven and dispersed worldly terrains might be crafted.

My effort to critically map reproduction's ontological politics is inspired by Ann Laura Stoler's suggestive and reflexive claim that several decades of recent critical scholarship in the humanities has produced a particular "regime of truth" on race, in which antiracist (and, one might say, also feminist) accounts of "race" start and end with the premise that "race" is socially constructed within nationalist, economic and colonial forms; "Race" is thus held as historically specific, and not fixed, even if it is a felt condition with material consequences.[3] This historicity and madeness of race can also be in turn historicized, understood as part of an ongoing struggle over the terms of existence of "race" as a differential condition in the world. That this chapter uses the phrase "sexed living being," for example, demonstrates how it already participates in an ontological politics, placing "sexed" as an adjective, a form of power-laden becoming, which materializes living being in historically specific ways. *Sexed*, like *raced*, designates difference as an evoked, nonoriginary modifier of living being.

Science and Technology Studies (STS)—an interdisciplinary conversation begun in the 1960s concerning the immanence of politics and culture in epistemological practices—has also been obsessed with the production of things-in-the-world out of the techniques, clinics, laboratories, models, and field sites of scientific endeavors.[4] Today, feminist STS scholars have shown that reproduction—as a process of living being—has not only been technologically altered but is caught in regimes that compel its malleability and generative capacities.[5] Through biomedical practices, reproduction in the contemporary moment can become what Charis Thompson calls a "Biotech mode," particularly at its micrological and clinical scales, in which tissue cultures, in vitro fertilization, PCR, genomics, virology, and bioengineering transform the generative lively capacities of the micrological substrates of cells, nuclei, DNA, proteins, and molecular processes into important forces of production for the creation of commodities, biovalue, and biocapital at the very same time that they constitute life.[6] Relatedly, since the 1970s, cheap mass-produced methods of contraception, sterilization, and abortion have offered cold war, postcolonial, and feminist projects alike techniques to redirect, alter, and preempt the fertility of millions around the world. Through assorted nationalist and transnational family planning projects, women in discrepant and diverse sites are nonetheless hailed similarly as embodied subjects whose fertility can and should be managed, even if more comprehensive infrastructures of health care are absent.

In these ways, in my lifetime, the dominant ontological status of human sexed living being, entangled as it is in technoscience and governmentalities, has reversed itself. Although feminist and antiracist STS scholars in the 1980s and 1990s tended to overturn scientific accounts of the natural givenness and fixity of sexed and raced bodily kinds, today they tend to track the artifice of sex—how sexed living being has retwisted a domain of alterability, imbued with the generative capacities of making and recombining, open to both fostering and preemption. At the same time that recombinatory capacities of life inaugurate new forms of capitalist value, they also open up the possibility of challenging and materially disassembling heterosexuality, racialized kinships, and sexual embodiment, opening life to queered reassembly. Thus, I want to think critically about the stakes of this current ontological politics of reproduction, in which it is celebrated for its malleability and changeability, in which the fostering of life is also at the same time understood as its alteration, management, and speculative preemption.

Feminists have played their part in conjuring this ontological politics of sex in the age of technoscience, having spent enormous energies showing how, in Judith Butler's formulation, the materialization of bodies as matter-prior-to-power is in fact power's greatest effect.[7] To this sense of sex and reproduction as made and malleable is often sutured the liberal feminist insistence on the female individual as the ethicized and responsibilized subject in relation to reproduction—compelling choices and acts at this individual embodied scale in the management of reproduction relative to biomedical possibilities. The infrastructures that support individualized choice might ethicize the female subject as responsible for reproduction and excise culpability of larger conditions of possibility. Overall, the two normative axes in the emergent ontological politics of bodily reproduction in the age of technoscience are, first, a malleable and managed embodiment *and,* second, an ethicized possessive individuality located in the property of the body.

Although continuing to insist on the material life-and-death consequences of reproduction, I consider here a figuration of reproduction as a *process* that exists at *macrological* (not merely micrological and bodily) registers and which is *extensive* geographically in space and historically in time. I will call this the distributed ontology of reproduction, or more succinctly, *distributed reproduction.*

Why distributed reproduction? "Reproduction" as a term has only in the last half century become apprehended as a phenomenon that happens primarily in bodies rather than at larger scales; thus, it may be fruitful to think suspiciously about this supposed common-sense site of the reproductive body and what it forecloses. A thread of "reproductive health" politics now entangles much of the world, manifest through NGOs, local clinics, and grounded projects as much as transnational organizations that offer grants and instructional guidelines. The emergence of "reproductive health" was fought for in the early 1990s by an emerging class of feminist development and family planning professionals who sought a more feminist-friendly way of articulating the need for health services, untangled from the coercive, sometimes violent, often misogynist, and deadly regimes of population control that had characterized much of the cold war/postcolonial period. The framing of reproductive health has tended to foster (not without debate) a politics of either improved health services or individualized reproductive rights and "empowerment."

Without negating the often life-enhancing effects of the work "reproductive health" has done, taking a step back as a historian and STS scholar raises questions about what condensing reproduction to an embodied and medicalized "reproductive health" forecloses. How might critical feminist health politics be reinvigorated by questioning how "reproduction" both includes and complexly exceeds the body, thereby mapping body as unevenly working through bodies in space and time?

More specifically, thinking through "distributed reproduction" builds on a set of less dominant, though still influential, feminist figurations of reproduction as entangled with political economy. There is an important itinerary of feminist, antiracist, and decolonizing critiques of capital that have excavated how reproduction *exceeds* the body through social reproduction, so that the politics of housing, war, immigration, labor, pollution, incarceration, and care giving are all profoundly also questions of reproduction; a list developed in the 1970s and to which today we might add politics of development, of biomedicine, of technoscience, of biocapital, of racial states reconfigured in the name of antiterrorism, of preemptive and privatized war machines, of pharmaceuticals that call for markets of perpetual risk, of affective economies, of planetary environmental crisis, of citizenship in terrains of displacements, of global capital's transnational disjunctures, and of biopolitics and necropolitics alike.[8]

Since the late 1980s, the Bangladeshi feminist Farida Akhter has offered trenchant critiques of both population control in its imperial and nationalist registers, as well as "reproductive rights" as its feminist alternative. In 1989, organized by Akhter, feminist activists and intellectuals primarily from South Asia and Europe wrote a declaration critiquing the ways cold war experiments had so thoroughly tied reproduction together with capitalism and technoscience—a knot they saw as facilitated by the ways liberal feminists had consolidated around a vision of a universalized female ethical subject who just needed her reproductive rights to do right. The Declaration of Comilla was important for the way it situated the politics of reproduction within the "engineering and industrialization of the life processes" more broadly.[9] In this declaration, reproduction stretched beyond bodies to implicate the multiple domains of industrialism and its environmental effects, family formations, agriculture, and the ownership of biodiversity, thereby necessitating a sweeping critique of both technoscience and capital. Their vision of an expansive reproductive politics was not remediable by the free choices of an individualized ethical subject.

Further, Shellee Colen's influential notion of "stratified reproduction" has aptly named how "kinship" is hierarchically rearranged by structures of race, sex, and class in transnational political economies.[10] Marilyn Strathern has helpfully developed a notion of "dispersed kinship" to ask complex questions about the shifting range of "procreaters" who take part in, "assist," and hence are in "relation to" reproductive acts as mediated by technoscience, property forms, and knowledge production.[11] The notion of "distributed reproduction" is kin to all these moves to apprehend reproduction as stratified and dispersed, as well as the project of "reproduction justice" fashioned by Asian Communities for Reproductive Justice and the SisterSong Collective in the United States, which builds on legacies of work by antiracist feminists and radically expands the sites where reproductive politics occurs to "issues such as sex trafficking, youth empowerment, family unification, educational justice, unsafe working conditions, domestic violence, discrimination of queer and transgendered communities, immigrant rights, environmental justice, and globalization."[12] This itinerary of critical work on "reproductive justice," stratified and dispersed kinship, and more provokes a need to rematerialize what counts as reproduction itself. It is precisely because bodies matter that there is urgency in tracking how bodies are discrepantly enrolled, altered, and announced as alterable by processes and structures that exceed the body proper.

For all these reasons, and more, I want to describe this distributed ontology of reproduction as composed of multiple *formations of reproduction* in a way that resonates with the phrase "formations of capital" as a description of historically specific relationships that produce and mobilize "capital." Fortunately for my project, there is a long history of understanding reproduction as a process of aggregate living being. "Reproduction" as a term that applies to biology emerged in the eighteenth century, amid the revolutionary installation of liberal politics that enshrined individuality and private property during the dawn of industrial capitalism. Reproduction has been simultaneously a term that describes the maintenance of *species being* and a term that describes the maintenance of the relations that constitute capital accumulation. Buffon, the famous French comparative anatomist, was the first to use the term "reproduction" to name a process of maintaining a species, a process that generates and maintains the stability of form and ways across generations.[13] Thus reproduction was at its inception an extensive process, binding and producing an aggregate living formation—the species—entangling, working through, and creating the individual organisms of that species. In other words, eighteenth-century "reproduction" was a process of replacement, sameness, and consistency that linked embodied individuals together as a common kind. Reproduction worked through bodies, but exceeded them.

Within the evolutionary epistemologies of the nineteenth century, "reproduction" continued as a species-level process of living being, yet one that occurs in extensive spans of time, constituting the ongoing historicity of life.[14] Moreover, reproduction as an evolutionary process did not simply maintain species-kind, but instead produced the *difference* that natural selection sorted. Reproduction was thus rendered a selective becoming-in-time that generated variation—a living difference engine. Reproduction stretched beyond mere mating, to occur at time scales larger than the life of the individual organism, into the recesses of evolutionary time. Already in the nineteenth century we can see the seeds of our current ontological politics, where reproduction becomes a process producing variety, difference, change, and innovation—an evolutionary process that by the early twentieth century was considered to be in need of rational guidance and engineering by technoscience, giving rise eventually to the progressive and genocidal projects of eugenics. Evolutionary macrological ontologies of reproduction are haunted by the selective deaths—by natural or engineered selection—that bring future life into new form. Therefore, the project of imagining a distributed ontology of reproduction is noninnocent, and it has its own deadly black holes in need of disavowal, making the retreat to the body all the more understandable.

This chapter hopes to reanimate this history of understanding reproduction as a distributed process that manifests in and connects, but is not reducible, to bodies. If reproduction is a distributed process of living being already transformed by birth control, biomedicine, biotechnology, infrastructures, pollution, housing, militarization, development, criminalization, nation-states, queer politics, labor-relations, and so on, what is an ontological politics of reproduction that can render legible how life is constituted through the infrastructures and political economies that exceed sexed and raced bodies as such? How to stretch attention to the temporally and spatially extensive matrixes of technoscience and political economy that do not just converge on, but are themselves the process(es) of reproduction?

In this endeavor, reproduction is reiterated in its initial double formulation as a process of biology and political economy. To insist on reproduction as a macrological *process* that conditions life is also to think harder about how "processes"—metabolic, chemical, biotechnical and engineering—have themselves become patentable, and thus amenable to proprietary formulation in commodity circuits. Thus, imagining distributed reproduction as an extensive process must be attentive to the evasive tactics of such property relations, but also draw them into constant critical legibility toward politicization.

In other words, if the conceptual goal of distributed reproduction is to provoke a different kind of "reproductive politics" than is commonly practiced, one critically implicated in the unevenly dispersed and spatialized relations in a world riven by capital flows, racialized geographies, wars and nation-states, then it is at least partially because I also want to highlight how relations of reproduction are entangled with, but not simply subsumable to, relations of production.[15] To give this account of distributed reproduction some empirical meat, I proceed by offering two entry points (among many possibilities) into mapping the formations of reproduction that make up its distributed ontology as it has congealed since the end of the cold war and in the aftermath of postcolonial projects of "modernity." The first is symptomatically represented by the figure of "human capital" crystallized in a process I call the *economization of life*, while the second takes the form of uneven saturations of chemical exposure.

A GIRL MAKES THE WORLD

At the 2009 World Economic Forum, a private nonprofit summit held annually at Davos, Switzerland, world leaders, economists, CEOs of Fortune 500 companies, supranational elites, and philanthropic luminaries met to discuss economic development in light of the urgent global recession. Amid the presentations on how to rescue global capital was a "popular" panel on the "Girl Effect." "The Girl Effect" is the name of a development campaign initiated by the Nike Foundation and Novo Foundation (affiliated with Warren Buffet's family) that argues the best way out of the world's current economic "mess" is to "invest in a girl and she will do the rest."[16] The girl will save the world because decades of social science data now calculate that if her education is increased, she will have fewer babies and she will have higher wages; Unlike her brothers, she will share her wages with her family and will more likely pay back loans; She will marry at an older age and have a higher dowry value.[17] As Lawrence Summers influentially declared in 1992 as chief economist of the World Bank, the girl is "the greatest investment of all."[18] For Nike, a corporation already successfully cathected to exuberant and individualized "Just Do It" advertising that draws on individualized desires to self-improvement, the ideology of "girl power" tethered to development made synergistic sense.

The girl in question is a historically specific subject figure. She is the longstanding sexed and racialized concern of development taken back to childhood. A world-enveloping social science infrastructure has for decades attended to the living being of the woman this girl-child will be/was born by. With ferocious intensity, beginning in the 1960s, the fertility of poor nonwhite women around the world was declared a ticking time bomb by US foreign policy, imperiling democracy, capitalism, and the planet. The population bomb tethered to feminist aspirations of the individual management of one's own fertility—the ambition of "seizing the means of reproduction" at the scale of the individual—gave rise to a proliferate international family planning industry.[19] For the first time in history, cheap mass-produced birth control and industrial-scale sterilization offered the technical means to materially govern human reproduction en masse, while tantalizingly promising a freedom from unwanted fertility for the individual. The question of human fertility, in turn, became pivotal to the postcolonial project of fostering and governing "the economy" as it emerged as the primary concern of state governmentalities. "The economy" is itself a historically recent entity, born in the twentieth century of macroeconomics and state data collection that offered new measures such as national unemployment rates, rates of inflation, and, crucially, gross domestic product (GDP) as the governable and alterable qualities of a nation's "economy."[20] Moreover, with decolonization, the planet was remapped through the universalized unit of the nation-state and hence reterritorialized as a landscape of governable economies connecting and differentiating regions of the planet into open and closed markets, developed and underdeveloped economies. Likewise, demographic data collection increasingly imbued national "population" with governable qualities: literacy rates, birth rates, mortality rates, but also measures of the desire to control fertility and measures of the availability of the commodities and services to do so.

The term *economization of life* names the historical emergence of forms of governmentality that sought to govern living being, particularly sexed-living-being and fertility—for the sake of fostering economic development and enhancing national GDP.[21] The economization of life names a historically specific formation organizing life, not for the purpose of generating surplus value or producing commodities, but for the sake of "the economy." Best known through population control and family planning, the governmentalities forming the economization of life grew out of racist eugenics logics that had flourished earlier in the twentieth century. With the economization of life, however, the project of intervening and altering reproduction was directed at enhancing *economic futures*, not racial evolutionary futures. While population control practices were typically still animated by racial-ized narratives of "civilized" and "underdeveloped," the explicit purpose of governing reproduction was for the sake of altering GDP, not purported racial destinies. Thus, the concept of the economization of life draws atten-tion to the significance of sexed living being to the rise of neoliberal govern-mentality in postcolonial itineraries.

Important to comparative measures of economic development was not only the measure of GDP (which varied widely between economies of dif-ferent sizes) but "GDP per capita," for which population growth has an enormous calculative effect. By the late 1960s, one RAND economist influ-entially calculated that money spent for each "averted birth" was "100 times more effective" in raising per capita GDP than the same amount spent on "productive investments" such as machinery or agriculture—a claim that helped to further spawn cost-benefit analyses for specific family planning interventions.[22] This calculus was crucial in directing US foreign aid funds to family planning over health, food, or kinds of aid, as well as making family planning crucial to President Johnson's domestic "war on poverty," making the United States not only the largest twentieth-century funder of population control, but also the most important global distributor of con-traception.[23] At the ideological heart of population control was the claim that reducing population growth rates was critical to creating futures of national productivity captured in the measure of GDP per capital. Hence, while eugenic necropolitics declared that some must die so that others may live more healthfully, with the economization of life this logic was translated into some must not be born so that future others might live more abun-dantly (consumptively).[24]

Though only sketched here, the economization of life is an important fea-ture of the last half-century, underwriting designations of lives more and less productive, lives unworthy of birth, and hence surplus life and "avertable" lives.[25] Thus, while formations of capital rely on the disposability of worker's lives—the cheapening of "labor power" to the point of its disposal—the economization of life names the extensive and discrepantly expressed forma-tion assembled out of governmentalities, family planning services, contra-ception, feminism, and social science infrastructures that harnessed birth to the fostering of GDP. In other words, it is a formation of reproduction.

Since the end of the cold war and the many-sited, many-flavored femi-nist interventions into the violent logics of population control, today the economization of life is increasingly transformed through the figure of "human capital." Human capital is a subject-figure of neoliberal economics that became in the 1990s central to current practices of development, war planning, education, and state building, supplanting population control and structural adjustment of the late-twentieth century with a purportedly more humanistic, empowering, feminist, and life-giving alternative. Human capital is defined as the knowledge, skills, values, health, and embodied capacities of people that make them economically productive. As human capital, every person is considered an entrepreneur in his or her own life.

When the World Economic Forum discusses "the girl effect," they are arguing for investment into a certain form of human capital embodied in girls. Crucially, and as Summers at the behest of the World Bank argued, girls' education is the best investment not only because it has good rates of returns for the girl, her family, and the state, but also because the same amount invested in education reduces her fertility more than it would if invested directly in family planning.

Here, I might observe the following points: (1) How fertility reduction has become so thoroughly associated with improved economic productivity that it now serves *as a proxy* measure for further removed correlations with economic productivity; (2) How reproductive interventions are temporally pushed forward in the human life cycle to the prechildbearing years of life;[26] (3) How the capacities of living being and the labor of social reproduction are now valued as forms of capital, placing reproduction and capital into a new tension.

The concept of human capital, first forged in the 1960s in terms of farming and fertility, was in the 1970s and 1980s most prominently deployed in human resource management interested in highly educated bourgeois "knowledge workers." It was only in the 1990s that human capital emerged as a hegemonic component of neoliberal development economics with the World Bank as its epicenter, joining with microcredit projects to herald "poor" racialized females as the most productive site of investment, debt, and entrepreneurialism.[27] In the process, the third-world girl has rapidly replaced the bourgeois knowledge worker as an iconic figure of human capital and the adult woman as the quintessential subject of development. Implied in the hypervaluation of educated girls is the devaluation of the adults that uncapitalized girls grow up to be as a future form of underproductive or even disposable life no longer worthy of investment, and the implicit devaluation of boys who offer lower rates of return.[28]

In sum, reknitting homo economicus as a form of capital embodied in young girls is at the same time a formation of reproduction, in which generative capacities of sexed living being are rearranged, revalued, and brought into new alignments and attachments. The girl effect, in the name of feminism and as a more humanist form of neoliberalism concerned with "human development," is congealing as the *liberal* neoliberal alternative to more brutal forms of neoliberalism associated with structural adjustment and disinvested infrastructures in the name of free markets. In the current era of

Barack Obama, racialized youth, born into sedimented dispossessions, are reinterpellated as a site of preemption and speculative investment that takes as its object a sexed moment in the individual's prereproductive life cycle. The girl effect is a hegemonic call to build a particular formation of reproduction, in which the burden of fixing capital in the face of its ravages are placed on the shoulders of a particular subject of sexed living being. Here, feminism and capital collapse into one another, and the project of feminism must face itself as an appropriated politics within this noninnocent formation of reproduction. Unlike the project of cyborg politics, the answer to the hypervalued girl is not a human-capital-politics that comes up with yet better ways of turning capital, feminism, and social science to the project of living, but instead reveals how that call—the call to make better feminist fixes—is no longer a subjugated critical epistemology, but constituted hegemonically in formations of reproduction. Distributed reproduction, in turn, passes through the World Bank, microloans, feminism, girls, and Nike sneakers.

Unsurprisingly, while the award-winning YouTube Girl Effect campaign brings tears to the eyes of some, others in the blogosphere are quick to point out that Nike is infamous for its history of exploitative labor practices, taking advantage of cheap, young, and female labor of global export zones. Geographer Melissa Wright shows how exports zones explicitly advertised young women's labor as cheap and disposable, that is as women not needed for the future reproduction of the nation and who themselves do not have futures, to attract the business of companies such as Nike.[29] What knits together the investable girl and the disposable girl are the matrixes that hail their future reproduction, their unconceived (let alone unborn) potential children, as undesired, devalued, and avertable.

CHEMICAL ALTERITY

This legible and celebrated figure of the hypervalued girl—a body whose improvement is desired even as her future reproduction is not—resides in juxtaposition with another formation of reproduction making up the tangle of distributed reproduction. This second example of a formation of reproduction moves the text from terrains of purported underdevelopment to Sarnia, Ontario, the "chemical alley" of Canada, a so-called developed country celebrated by (some) Americans for its social welfare infrastructure and peace keeping.[30] Here, the politics of distributed reproduction is not hypervalued, but rendered externalized and uncertain.

With thirty-five major petrochemical, polymer, and chemical factories, Sarnia is one of many concentrated nodes of intensive petroproduction, attracting companies with reduced environmental regulations and tax breaks, hooking up the planet to oil in snakes of pipelines, tankers, and refineries, bringing "good things to life" and unevenly saturating the world with poisonous and climate altering effluvia. Sandwiched amid this concentration of petroproduction is the Aamjiwnaang first nation, where environmental justice activists have worked with local scientists, feminists, and lawyers to document the first-known case of a dramatic reduction in the birth of boys associated with chronic chemical exposures.[31] Today, only thirty-five boys are born for every hundred girls.[32]

In this second example of a formation of reproduction, I want to attend to the effects on living being that are abjected from economic calculi—the modifications to living being and its reproduction that are ubiquitous and intensifying even as they have become invisibilized or externalized from hegemonic economic regimes of value. Pollution, as the excess of production, an excess turned away from by its creators and by the nation-state, is nonetheless a material and altering presence in living being.

In this formation of reproduction, pervasive, mobile, and accumulated chemicals have saturated nonhuman and human living being with modifying effects to sexed living being at increasingly noticeable levels. Such modifications are embodied individually, with painful and tragic tolls on lives, but at the same time are distributive across individuals (human and nonhuman) into a shared, yet unevenly dispersed, condition of having been already altered. Chemical injury, produced by industrial production or later by commodities, not only causes cancers or poisoning, but alter the material substrates of reproduction, mimicking or disrupting hormonal signals, mutating genomes, and thinning membranes. If living being is now hailed as alterable, and materially transformable in new ways, opening new possibilities for a malleable ontology of life, chemical injury calls for a more critical politics of alterability and greater attention to the kinds, modes, and exercise of power manifest in malleable life.

If temporal displacement is an important feature of the economization of life (in which living being is intervened in now to prevent reproduction later), displacement is also at work in the violent effects of externalized chemical excess. Chemical injury is not just displaced *spatially* with super stacks, toxic trading, and selective incinerator placement to funnel its effects to less-regulated zones, to disenfranchised locations where bodies can be rendered more disposable, but also displaced *temporally*, such that accountabilities exceed the scope of individual lives—accumulating or persisting over time. The possible effects are not only felt at the moment of the exposed organism, not only in the future of potential lives yet to be born, but also in the future generation of possible grandchildren. Research into the effects of the estrogenic chemical bisphenol A (BPA) on pregnant mice has found that the significant effects occur not so much for the fetus in utero, but for the eggs being formed inside that fetus, and hence effects are manifest for the potential grandchildren who will not be born.[33] Here, the lives not born in Sarnia may well be the effect of exposures endured by their grandmothers.

Although the violence of chemical injury is concentrated in the space of Aamjiwnaang land, a parcel made available to such injury by a colonial history of displacement and racist dehumanizing of first nation people, chemical injury is ever-present as what S. Lochlann Jain calls "commodity violence," in which the harmful effect of commodities are probabilistic for the population, but not causally isolatable or predictable at the individual level.[34] Thus, it is ironic—or rather constitutive—that when the Canadian NGO Environmental Defense performed a national biomonitoring study called "Toxic Nation," including not only the family of an Aamjiwnaang environmental activist but also a handful of high-ranking politicians, it was the politicians who proved to have the overall highest concentrations of tested chemicals, more so than any of the citizen test subjects.[35] Is it ironic or constitutive that Lawrence Summers wrote not only the pivotal 1992 document that declared the girl the greatest investment, but also an infamous World Bank memo in 1991 that sarcastically suggested exporting toxic industries to less-developed countries, because "the economic logic behind dumping a load of toxic waste in the lowest wage country is impeccable and we should face up to that."[36]

As already chemically saturated beings, already changed and not just potentially or intentionally malleable, the ontological politics of reproduction turns on this problem of alterability. The politics of alterability is not just a question of what change to choose. It is also a politics of the differential availability of alteration to stratified subjects, alterations that can be unwanted and not just wanted, that can be life-taking and not just life-fostering. Alterations that not only happen in bodies, but also bind and dislocate bodies across time and space. This connection of industrial production with living being via chemicals is easily politicized as a moral panic over "missing boys" and nonheteronormative bodies. Phrasing this problem as one of "missing boys" tends to uncritically grant "boyness" and "girlness" as an a priori status of the fetus, which is then altered into intersexuality or even death, reiterating (in the name of environmental critique) naturalizations of biological sexual difference.[37] Rethinking endocrine disrupters and other body-altering chemicals as constituent elements of a distributed ontology of reproduction hopes to invite much needed collaborations between feminist technoscience studies, transgender and queer studies scholars, and environmentalists to offer better accounts of the uneven political stakes of alterable living being.[38]

CHANGING THE POLITICS OF CHANGE

As two formations that feature "change," the politics of human capital and the politics of chemical injury become entwined. Discrepant formations of reproduction intersect to become a complex and contradictory distributed ontology, which in turn demands new political imaginaries. Although feminist have long been predicting the coming world of reproductive technologies, is feminism sufficient to the task of this more multidimensional ontological struggle over living being? What would it look like to craft an account of the distributed ontology of reproduction that shows the possible connective tissue formed through, for example, human capital, chemical injury, the biotech mode of reproduction, and transnational family planning? Reproduction is conjured as an extensive assemblage of formations, drawing divergent domains into its ambit—a process that not only works through embodied lives, but also operates at time scales larger than the human, demanding collective effort.

If my question is how to imagine and politicize distributed reproduction, then the scale of this project is so enormous that I stagger before it as a single person, and even the question dramatically escapes the scope of this chapter. If distributed reproduction can be characterized, at a first approximation, as unevenly extensive in space and time, necessitating both topological imaginaries and temporal politics, it generates such uneven conditions as much as builds on them. Distributed reproduction is constituted out of technoscience, nation-states, capital, and infrastructures as much as out of bodies, living being, sex, and race. The "scattered hegemonies" of distributed reproduction need to be critically traced as much as the alterations to living being differentially politicized.[39] Distributed reproduction is noninnocent, and hence it is not just a domain of life creation but a processional exercise of power that is as necropolitical as it is biopolitical. At the same time, distributed reproduction is binding and attaching, and thus holds open unexpected possibilities for new forms of kinship and alliance across space and time.

It has become a cliché to feel that one is living through a moment of tremendous change. If humans are undergoing a rearrangement of sexed living being, for which this chapter is a symptom as much as a diagnosis, then the feminist dream of valuing social (and cultural) reproduction is turning monstrous. Formations of reproduction herald sexed and raced life as the greatest investment at the same time that other entangled formations of reproduction are produced out of the indifference of capital to its material effects on organisms, with deadly repercussions. Sexuality too is under rearrangement, with the vital stakes of queer desire entangled in capital's call. What is the political imaginary that might name this skein?

NOTES

1. Stacy Leigh Pigg, "Globalizing the Facts of Life," in *Sex in Development: Science, Sexuality and Morality in Global Perspective*, eds. Stacy Leigh Pigg and Vincanne Adams (Durham, NC: Duke University Press, 2005) 39–65; C. P. Macpherson, *The Political Theory of Possessive Individualism* (Oxford: Oxford University Press, 1962).
2. Donna Haraway, " Manifesto for Cyborgs: Science, Technology and Socialist Feminism in the 1980s," *Socialist Review* 80 (1985) 65–108.
3. Ann Laura Stoler, "Racial Histories and Their Regimes of Truth," *Political Power and Social Theory* 11 (1997) 183–206.
4. Ian Hacking, *Historical Ontology* (Cambridge, MA: Harvard University Press, 2002).
5. See for example, Sarah Franklin, *Dolly Mixtures* (Durham, NC: Duke University Press, 2006), Michelle Murphy, *Seizing the Means of Reproduction* (Durham, NC: Duke University Press, forthcoming), Catherine Waldby and Melinda Cooper, "The Biopolitics of Reproduction: Post-Fordist Biotechnology and Women's Clinical Labour," *Australian Feminist Studies* 23, no. 55 (2008) 57–73.
6. Charis Thompson, *Making Parents: The Ontological Choreography of Reproductive Technologies* (Cambridge, MA: MIT Press, 2005).
7. Judith Butler, *Bodies that Matter: On the Discursive Limits of "Sex"* (New York: Routledge, 1993).
8. For an example of this more capacious sense of antiracist reproductive politics see African American Women for Reproductive Freedom, "We Remember," in *Still Lifting, Still Climbing: African American Women's Contemporary Activism*, ed. Kimberly Springer (New York: New York University Press, 1999). On necropolitics, see Achille Mbembe, "Necropolitics," *Public Culture* 15, no. 1 (2003) 11–40.
9. "Declaration of Comilla." *Journal of Issues in Reproductive and Genetic Engineering* 4, no. 1 (1991): 73–74.
10. Shellee Colen, "Like a Mother to Them: Stratified Reproduction and West Indian Childcare Workers and Employers in New York," in *Conceiving the New World Order: The Global Politics of Reproduction*, ed. Faye Ginsburg and Rayna Rapp (Berkeley: University of California Press, 1995) 78–102.
11. Marilyn Strathern, *Kinship, Law and the Unexpected: Relatives Are Always a Surprise.* (Cambridge: Cambridge University Press, 2005).
12. Sister Song Women of Color Reproductive Health Collective, *Reproductive Justice Briefing Book (Atlanta: Sister Song, 2007)* and Asian Communities for Reproductive Justice, A *New Vision for Advancing our Movement for Reproductive Health, Reproductive Rights, and Reproductive Justice* (Oakland: ACRJ, 2005).
13. Jacques Roger, *Buffon: A Life in Natural History*, trans. Sarah Bonnefoi (Ithaca, NY: Cornell University Press, 1997).
14. Staffan Mueller-Wille, "Figures of Inheritance, 1650–1850," in *Heredity Produced: At the Crossroads of Biology, Politics, and Culture, 1500–1870*, ed.

Staffan Mueller-Wille and Hans-Joerg Rheinberger (Cambridge, MA: MIT Press, 2007) and Elizabeth Grosz, "Darwin and the Ontology of Life," in her *Time Travels* (Durham, NC: Duke University Press, 2005) 35–42.

15. Haraway's "Cyborg Manifesto" is often contextualized as a science-positive politic relative to debates in the 1980s about feminism's repudiations of "patriarchal rationality," thereby typically passing over the fact that cyborg was a figure that written relative to Marxist feminism as much as poststructural or posthumanist politics.

16. Text from Nike Foundation The Girl Effect campaign video. http://www.girleffect.org/

17. Such correlations are listed in the "factsheet" of the Nike campaign, but are also endemic in the field of development studies. http://www.girleffect.org/#/fact_sheet/

18. Lawrence Summers, "Investing in *All* the People," in *Policy Research Working Papers, World Bank* (1992), Lawrence Summers, "The Most Influential Investment," *Scientific American* 267, no. 2 (1992) 132.

19. Murphy, *Seizing the Means of Reproduction*.

20. Suzanne Bergeron, *Fragments of Development: Nation, Gender, and the Space of Modernity* (Ann Arbor: University of Michigan Press, 2004); Timothy Mitchell, "The Work of Economics: How a Discipline Makes Its World," *European Journal of Sociology* 47, no. 2 (2005) 297–320.

21. For a detailed history of the economization of life, see Murphy, *Seizing the Means of Reproduction*.

22. Stephen Enke, "Birth Control for Economic Development," *Science* 164, no. 388 (1968); Stephen Enke, "The Economic Aspects of Slowing Population Growth," *Economic Journal* 76, no. 301 (1966): 44–56.

23. Matthew Connelly, *Fatal Misconception: The Struggle to Control World Population* (Cambridge, MA: Harvard University Press, 2008); Lyndon B. Johnson, "Address in San Francisco at the 20th Anniversary Commemorative Session of the United States," (1965).

24. The eugenic logic described here is a paraphrase of Foucault's formula for the racial state. Michel Foucault, *"Society Must Be Defended"*: *Lectures at the Collège de France* 1975–76, trans. David Macey (New York: Picador, 2003).

25. "Births averted" was a common population studies measure.

26. On the temporal and anticipatory politics, see Vincanne Adams, Adele Clarke, and Michelle Murphy, "Anticipation: Technoscience, Life, Affect, Temporality," *Subjectivities* 28, no. 1 (2009): 246–265.

27. Lamia Karim, "Demystifying Micro-Credit: The Grameen Bank, NGOs, and Neoliberalism in Bangladesh," *Cultural Dynamics* 20, no. 1 (2008) 5–29.; Katherine Rankin, "Governing Development: Neoliberalism, Microcredit, and Rational Economic Woman," *Economy and Society* 30, no. 1 (2001) 18–37.

28. Geeta Patel's work on "financial selves" and "risky subjects" elaborates on this mapping of speculative risk onto life, Geeta Patel, "Risky Subjects: Insurance, Sexuality, and Capital," *Social Text* 24, no. 4 (2006) 25–65.

29. See parody video response by the contrarian organization Corrupt. http://
    www.youtube.com/watch?v=x_lLdYb2z1g; Melissa Wright, *Disposable
    Women and Other Myths of Global Capitalism* (New York: Routledge,
    2006).
30. For a more detailed discussion of Sarnia, see Michelle Murphy, "Chemical
    Regimes of Living," *Environmental History Journal* (2009) 695–703. For a
    detailed analysis of another North American "chemical alley" see Barbara
    Allen, *Uneasy Alchemy: Citizens and Experts in Louisiana's Chemical Corridor
    Disputes* (Cambridge, MA: MIT Press, 2003).
31. See the environmental justice work of Ada Lockeridge and Ron Plain; Dayna
    Scott, "Confronting Chronic Pollution: Environmental Justice for the
    Aamjiwnaang First Nation," paper delivered at the University of Toronto,
    Jan. 2009; and the film *The Beloved Community* by Pamela Calvert/Plain
    Speech (California Newsreel, 2007).
32. Constanze Mackenzie, Ada Lockridge, and Margaret Keith, "Declining Sex
    Ratio in a First Nation Community," *Environmental Health Perspectives* 113,
    no. 10 (2005) 1295–1298.
33. Martha Susiarjo, Terry Hassold, Edward Freeman, and Patricia Hunt,
    "Bisphenol A Exposure in Utero Disrupts Early Oogenesis in the Mouse,"
    PLoS Genetics 3, no. 1 (2007) e5.
34. Sarah Lochlann Jain, *Commodity Violence* (Durham, NC: Duke University
    Press, forthcoming).
35. Environmental Defense, "Toxic Nation on Parliament Hill: A Report in
    Four Canadian Politicians" (Toronto: Environmental Defense: 2007).
36. Summers claims that this internal memo was meant as a sarcastic text meant
    to refine debate at the World Bank. However, economist commentators have
    pointed out that the argument of the memo, from an economic rationality,
    is consistent. For a defense of Summers' argument, see Jay Johnson, Gary
    Pecquet, and Leon Taylor, "Potential Gains from Trade in Dirty Industries:
    Revisiting Lawrence Summers Memo," *Cato Journal* 27, no. 3 (2007)
    397–410.
37. For a critique of the tendency to naturalize male and female sexes in
    debates over endocrine disrupters, see Celia Roberts, "Drowning in a Sea of
    Estrogens: Sex Hormones, Sexual Reproduction and Sex," *Sexualities* 6, no.
    2 (2003) 97–115.
38. For some starts in this direction, see Sarah Lochlann Jain, "Cancer Butch,"
    Cultural Anthropology 22, no. 4 (2007) 501–538.
39. Inderpal Grewal and Caren Kaplan, eds., *Scattered Hegemonies: Postmodernity
    and Transnational Feminist Practices* (Minneapolis: University of Minnesota
    Press, 1994).

CHAPTER 3

MULTICULTURALIST WHITE SUPREMACY AND THE SUBSTRUCTURE OF THE BODY

DYLAN RODRÍGUEZ

INTRODUCTION: FROM "RACE" TO WHITE SUPREMACY

The dreadful genius of the "postracial"/"postracist" moment lies in the creative disruption of white bodily monopoly in the operative sites of US nation-building (racial empire) from the grass roots to the White House. The ascendancy of postapartheid and "postcivil rights"[1] multiculturalisms marks the obsolescence of "classical" white supremacy as a model of oppression and socially ordering violence based *primarily* or even *predominantly* on the (relatively) exclusive vesting of hegemonic institutional power in the collective white social body. Postracial, postracist Americanism—accumulating momentum as the *still racial* nationalist narrative of the twenty-first-century United States—is far worse than a naïve or misinformed mythification of the civil rights dream: it is the signaling of a sophisticated, flexible, and "diverse" (multiculturalist) white supremacy as the heartbeat of the US national form.

The signature racial institutional shifts of the post-civil rights period have been marked by visible changes—compulsory and voluntary—in the comportment, culture, and workforce of white supremacist institutions: selective elements of police and military forces, global corporations, and major research universities are diversely colored, while their marching orders continue to mobilize the familiar labors of death- and misery-making (arrest and justifiable homicide, fatal peacekeeping, overfunded weapons research, etc.). Thus, while the *phenotype* of white supremacy has changed—a reordering (and partial "reprofiling") of bodies altogether necessary for its technologies of institutional power to remain viable under changed historical conditions—its internal coherence *as a logic of social formation* has been sustained, redeemed, and enhanced.

The allegedly postracial, postracist United States is a laboratory for the invention and refinement of global white supremacy, a production of institutionalized dominance and violence that inspires, hails, authorizes, and empowers its historical objects of vulnerability and dehumanization. As the systemic fatalities of "race" remain indelibly imprinted on the constituting social and cultural structures of state, nation, community, and public, the *dislocation* of fatal agency from the exclusive domain of the white body compels a transformation of activist and scholarly praxis. How are we to conceptualize, confront, oppose, creatively disrupt, and/or transform this multiculturalist renaissance of a global white supremacist form? How does the recasting and redistribution of racial/racist "bodies" within the *desegregated* geographies of white supremacy compel a reframing of antiracist, abolitionist, antigenocide, decolonizing, and radical liberationist work?

It is by now widely acknowledged that a critical scholarship of "the body" must substantively address "race" as a basic category of analysis, conceptualization, and theorization. Yet to address race and the body as such, that is, to emphasize how bodies (human, nonhuman, "virtual," and otherwise) are formed through the discursive regimes of racialization across time and place, is to already provoke sweeping historical-political questions about how power, violence, and domination *work on and through* particular kinds of (racial) bodies.

In this chapter, I want to highlight the urgency of embracing and engaging an emergent "turn" in critical interdisciplinary approaches to the body, a methodological and theoretical move that explicitly departs from *descriptive* approaches to race (which tend to rely on race as a taken-for-granted category of analysis and/or comparison) and instead focuses centrally on what might be termed the *racial determinations* of the historical-social form. Elaborating and stretching Cedric Robinson's durable conception of "racial capitalism,"[2] I am interested in an analytical schema that strategically privileges the technologies and socially constituting logics of power that constitute "race" as the organizing discursive structure of a social determination— white supremacy—that is relatively symbiotic with (*and not derivative of*) the economic substructure (and the "mode of production" generally).[3] This conceptualization suggests the following: (1) The contemporary moment of "multiculturalist white supremacy" is a *historically specific social formation* of a white supremacist substructure that is several centuries in the making and not the harbinger of white supremacy's (or race and racisms') extinction;

(2) This white supremacist substructure has, across historical periods and different social formations, worked *on* and *through* the body as both its primary social-discursive "raw material" and its main institutional mediation of the social, the cultural, and the political; and (3) unlike the historical capitalist substructure, the schematic logics of white supremacy are not accumulation, surplus value, and labor exploitation, but are *civilization* (read in verb, not noun form), *genocide*, and *incarceration*.

To paraphrase Stuart Hall, if race is not reducible to "class" relations (or for that matter, the economic substructure of a given social formation), but is *at the very least* the embodied social experience "through which class is lived," and "the medium in which class relations are experienced,"[4] then it is incumbent on critical scholars and activists to develop a useful framework through which to understand the structuring logics of race as a relation of dominance, and not simply as a socially produced (and hence epiphenomenal) category. To push at Hall's formulation a bit, not only is race *not* "*epiphenomenal*" to social (or "class") formation, it is *a primary determinant* of it.

Foregrounding the *material historicity of the body*—its contested physiology, social agency, historicity, and nonuniformity—is an essential prerequisite of such a framing: it is the raw historical material on and through which the discursive productions of "race" work and is the primary medium of the white supremacist substructure's social technologies. The body is at once a *discursive site, "material" mobilization, and biopolitical production*[5] of white supremacy's social technologies. It is caught up in the intense traffic of discourse and mobilization, representation and biopolitical power, "profiling" and historical agency. In what follows, I argue for a critical scholarship of race and the body that pivots on first, a historicized conception of white supremacy as the historical substructure of "race" and the ensemble of social relations within which race is embedded and implicated; and second, a dynamic understanding of how the body, in its flesh-and-blood concreteness as well as its conceptual/ideological slipperiness, is constantly altered by (as it circulates within) the changing structural logics of domination and violence that characterize white supremacy as a durable and flexible substructure of social formation.

I offer this meditation in the context of what seems to be a decisive conjunctural shift in the structural and institutional form, if not the empirical social outcomes, of racism and white supremacy in the "Obama moment" of US national formation and its persistent attempts to reproduce and retain a semblance of global hegemony. While the rhetoric of the "postracial" or "postracist" United States has failed (so far) to obtain sufficient ideological traction to comprehensively and decisively transform the racial common sense, this rhetoric's very discursive presence and cultural circulation demands a rigorous, scholarly, political-intellectual response that adequately negotiates the possibility that existing ways of talking and thinking about race and power—in academic, activist, and lay circles—may well be in obsolescence. We are left with a significant challenge: how can we explain the indelible presence of racism as "the state-sanctioned and/or extralegal production and exploitation of group differentiated vulnerabilities to premature death,"[6] while also accounting for the apparent *desegregation of white supremacy*, such that its embodied historical agency, as well as its institutional embodiments, can no longer (if they ever could) be conflated with the "white body" as such? When "people of color" become selectively, unevenly incorporated and engaged in the creation, leadership, and everyday operation of institutional racisms—as administrators, state officials, pedagogues, cultural producers, bureaucrats, ideologues, and armed executors—the conditions of possibility for "racist" *and* "antiracist" social agency have been structurally reformed. Thus, what sense can be made of the lasting social productions and systemic processes of the white supremacist substructure when the *bodies it enables, empowers, inspires, and mobilizes* include those that have been—and continue to be—targeted for enslavement, colonization, civil disfranchisement, militarized subjection, policing, and massive imprisonment?

"SUBSTRUCTURE"

I proceed with a conception of white supremacy as an internally complex, historically dynamic logic of social determination. In this sense white supremacy is related to, sometimes symbiotic with, but still analytically distinguishable from the more commonly accepted social determinations, "substructures," and macrostructural iterations of capitalism, empire, and patriarchy. What might it mean to engage the political and intellectual implications of conceptualizing white supremacy as a *changing* social form and *complex* historical logic that is both *traceable* through a historical genealogy and *identifiable* as a constitutive dimension of our present tense? In fact, this conceptualization and theoretical complication of race, power, and sociality can provoke precisely the critical research agendas and activist praxis required by what appears to be a qualitative if not monumental turn in the social formation of "race" and its still proliferating fatalities. For example: How do the emerging global and nationalist narratives of the "postracial" moment attempt to *address and/or displace* the particular social and cultural formations produced through the systemic logics of late-twentieth- and early-twenty-first-century racisms (e.g., Hurricane Katrina and the Ninth Ward, the protogenocidal Israeli occupation of Palestine, the global US prison regime)? Does the analytical rubric of "racialization" adequately or accurately encompass the discursive "mode of production" of the racial body, given the reformation of white supremacy in and through multiculturalism? How can liberal, progressive, and radical "antiracisms" articulate as practices of reform, social justice, and transformation when the apparent bodily signifiers of racism and white supremacy are substantively delinked from white social subjectivity and the white body?

The intersecting political genealogies of white supremacy (which can be traced across the knowledge productions of religious and theological discourse, political philosophy, canonical Western literatures, hegemonic racial modes of economic organization, etc.)[7] convene in multiple, apparently contradictory, or politically inverted social forms, from racial chattel slavery and formal apartheid to "post-civil rights" neoconservatism and multiculturalist Obama liberalism. The complexity and multiplicity of white supremacist social forms, in this sense, necessitates a critical scholarship of the body that does not reduce "race" to the processes of "racialization"[8] or even "racial formation."[9]

Rather, to position the "racial" body within the structuring political and social *determinations* of white supremacy invokes massive questions about the conditions of possibility for the discursive and physiological *production* of the body in its conceptual and flesh-and-blood reality. The gendered white supremacist substructure of racial chattel slavery, for example, has specific implications for how family genealogies, vulnerabilities to racist sexual violence, and marginalization within the jurisprudence of "miscegenation"[10] shape and reshape the racial body;[11] put differently, racial chattel slavery's white supremacist determinations have affected the historical condition of the black body most directly and violently, but have *also* partly composed the conditions of possibility for the white body, the indigenous body, the Asian body, and so on as we have come to know and inhabit them. Thus, while racially constituted bodies are formed through specific ensembles of social relations, discursive/cultural structures, and historical conjunctures (e.g., New World migration, land conquest, transpacific labor migration), their *determinations* are not isolatable from one another and can in fact be traced *relationally* across the changing systemic/institutional logics and cultural productions of a white supremacist substructure.

Although in one sense the social-discursive structure of "race" may be understood as an epiphenomenon of a particular and fundamental political-economic "base," or as a reflection of a substructure in which the economic mode of production remains the primary engine of social formation, the socially determining logic of white supremacy is both constitutive of this economic base/substructure and a relatively "independent" or autonomous substructural determination in and of itself: capitalism, empire, and so on are constitutively and productively white supremacist, and vice versa. There are historical moments, in other words, when the overlapping though relatively autonomous *determinations* of white supremacy and capitalism are in resolute tension or contradiction, such that they are not mutually *reducible* to one other; conversely, there are times when white supremacy and capitalism appear to exist symbiotically, such that their determinations seem to converge and coalesce with relative seamlessness (although it is also in these moments when it is most tempting to reduce white supremacy to an epiphenomenal effect of capitalism). Thus, to critically engage the intersection of race and the body without an essential historical *and theoretical* attention to how this intersection is enabled, transformed, and (re)produced through its structuring white supremacist genealogies is (at the very least) a profoundly ahistorical and subtly depoliticizing labor. I proceed from a partly polemical position: white supremacy cannot be justifiably positioned as a past tense political artifact, extremist ideological articulation, *irrational* enunciation of "hate," or reducibly essentialist or chauvinistic racial discourse.

Rather, in initially conceptualizing white supremacy as a productive logic of social organization, I am also centering it as a kind of *historical* substructure as well: white supremacy is no less a dynamic narrative *telos* of power and dominance that has shaped the discursive and political (i.e., "material") parameters of historicity itself, animating epochal articulations as far flung as the Treaty of Tordesillas (1494) to the current War on Terror. This is to suggest, as well, that the social-historical logic of white supremacy encompasses "preracial" and "protoracial" articulations of human hierarchy and difference, particularly within the pragmatic rubrics of conquest, genocidal war, and colonization. In this broad historical scope, the challenge for a critical praxis is to build a *theoretical* approach that (I hope) productively disrupts the empiricist tendency to neatly categorize white supremacy within a discrete and peculiar set of ideological, political, and institutional practices. To address white supremacy as a social-historical form, in this sense, is to attempt to explain the tensions and contradictions—as well as the congruencies and symbioses—between its substructural "determinations" of the social formation (which persist in different forms across historical periods) and its multiple constructions and inhabitations of "the body" across scales of institutionality, physiology, and phenotype. I am not dismissing the importance of the "empirical" realm in defining a social-historical form; to the contrary, I am arguing for a different epistemological premise and theoretical approach to apprehending (and thus *engaging*) white supremacy and its living archive. In the rest of this chapter, I will elaborate the terms of this critical engagement through a series of reflections on key terms—multiculturalism, determination, reform, phenotype—that depict a contemporary genealogy of white supremacy. While the central examples with which I am illustrating my conceptualization involve the bodies of the racial/racist state (from police to president), this privileging of white supremacist statecraft is meant as a provocation toward other analytical foci and not as an assertion of theoretical or political primacy. Finally, I conclude my thoughts with a somewhat nonprescriptive framing of the possibilities for a radical praxis or "agency" within the terms of *abolition*.

EROSION OF THE PROTOTYPE

The bodies of genocidal and protogenocidal racist violence, the fleshy materiality of modern and postmodern "whiteness" as a module of social formation, can no longer be adequately conceived and theorized through either analogical or analytical references to classical models of apartheid, segregation, racist chattel slavery, colonial conquest, or even "neocolonial" rule. Here I am most concerned with addressing what I believe is a generally misconceptualized and egregiously undertheorized turn in the trajectory of local and global racial formations, a shift that may render many of our most durable scholarly and activist analyses of race, racism, and (endemically racial) power obsolete, and which will certainly require a substantive transformation of antiracist, anticolonial, and antigenocide praxis.

The erosion of the "white body" as the singular prototype of racist and white supremacist agency—as the peerless and exclusive racial, physiological, and phenotypic template through which technologies of racial dominance and the white supremacist substructure's social-historical mode of production calibrate their institutionalized violences—disrupts both the "whiteness" and the "body" generally presumed to characterize and inhabit white supremacy. This is not to concede to the tepid liberalism of "postracial" and "postracist" political desire, nor is it to authorize cynical attempts to install "reverse" racism or "antiwhiteness" as the signature racial dynamics of our moment. Rather, it is to invoke the political *and bodily* desegregation of white supremacy as a change in the institutional and phenotypic composition of the racial substructure. Here, we might conceptualize *multiculturalist* white supremacy more rigorously and densely: as a diversity of *phenotypic* racial bodies variously (ideologically and culturally) produced, (pragmatically and vocationally) trained, and (circumstantially and militarily) coerced into a modality of social organization and cultural banality that *extrapolates* (and strategically *amplifies*) the violent technologies of classical (apartheid, white phenotype monopolistic) white supremacy. Within this racial social form, the question of whether and how these diverse racial bodies obtain or sustain a status of social "subjectivity" is structurally *secondary* and frequently irrelevant to the matters at hand; what matters most is the institutional location and discursive production of racial bodies in and of themselves and the capacity/willingness of these bodies to *perform* a social/ institutional "agency" within the logics of a desegregated white supremacy.

Stated differently, the emergence of a postapartheid state (in the United States and elsewhere) and the *bodily* racial desegregation of certain institutional spaces (government, military, state services, neighborhood, school, and beyond) have broken ground for white supremacy's flexible genius. Its capacity to sustain, transform, and/or elaborate the *global* social logics of racist subjection—criminalization, state-sanctioned bodily punishment, militarized vulnerability, and culturally normalized premature (violent) death—while claiming the moral-political high ground of multiculturalist diversity (postapartheid/desegregation) is one of white supremacy's landmark innovations, rather than evidence of its long-awaited social obsolescence. White supremacy is not reducible to a *singular* blueprint or paradigm of socially organized dominance—that is, "racial dictatorship,"[12] white institutional monopoly, or (proto)genocidal violence—but is a dynamic and flexible *social determination*, produced and affected by historical conditions and political relations of global, local, and intimate (bodily) power. In a sense, it reaches its highest point of social and institutional articulation when it has been "liberated" from the conventions of the apartheid logic and manifests within the normative political and cultural structures of (postracial) democracy, liberal humanism, and (national) progressivism. This is how the white supremacist substructure reforms, rearticulates, and potentially amplifies its determinations of social-historical formations—such as the post-civil rights, postracist/racial United States—that allege to have transcended or redressed the white supremacist denouements of racial colonialism, chattel slavery, genocidal conquest, apartheid, and land displacement.

Contrary to the most ambitious and radical political visions of antiracist, liberationist, and abolitionist praxis, this particular transformation in the embodiment and institutionality of racial power does not undermine or extinguish racism as the primary technology of the white supremacist substructure, but seems to multiply its possibilities of bodily inhabitation and institutional-intellectual leadership. Further, the white supremacist substructure is itself neither critically "exposed" nor (significantly) politically disrupted in this conjunctural moment, but seems to be rendered more flexible and adaptable in its capacities for organizing racially formed vulnerabilities to violence and fatality.

MULTICULTURALIST POLICING

The social logics of militarization, criminalization, and patrolling have been historically central to the construction, reproduction, and institutional coherence of white supremacist social formations. This centrality has been most evident in what I have been referencing as white supremacy's "classical" socialities: racial slavery, racial colonialism, genocidal conquest, and apartheid. The current moment, however, calls for a shift of conceptualization and theoretical approach that *sees* systemic white supremacist surveillance, apprehension, and systemic (state) violence somewhat differently. Although the different trajectories of "multiculturalism" as a critical intervention on eurocentrism, white institutional monopoly, and white nationalism have undoubtedly broadened the political terrain on which progressive and radical antiracist praxis can crystallize and mobilize, these have also been underestimated as a *potential site of inhabitation for white supremacist social logics and political technologies.* For example, when the Los Angeles Police Department (LAPD) undertakes vigorous recruitment efforts to increase the racial and gender "diversity" of its officers, sponsors the Gay Games VII Sports and Cultural Festival (held in 2005) as part of a national attempt to increase numbers of (out) gay cadets,[13] or cosponsors the tenth Annual International Criminal Justice Diversity Symposium (2006)[14] under the leadership of Chief William Bratton (the former New York Police Department [NYPD] commissioner notorious for authorizing citywide racist police harassment of blacks and Latinos under the rubric of "broken windows" policing),[15] its purpose and outcome is not to minimize or reduce the surveillance, arrest, and incarceration of (racially) criminalized populations or to mobilize a diverse police force to increase its enforcement of drug and alcohol offenses in high-income areas of Los Angeles, but rather to facilitate a multiplication of *white supremacist phenotype.* To render more complex what appears to be transparently obvious, the LAPD is not recruiting and hiring a more "multicultural" or diverse officer corps—through efforts like its recent "diversity awareness" partnership with the website Careerbuilder.com[16]—due to either a sudden transformation of its on-the-ground-policing ideology or a civil libertarian (and antiracist) commitment to ceasing its focused and strategic militarizations against (queer and transgender) sex workers, the homeless, undocumented workers, and poor black and brown youth. Rather, the LAPD is invested in building a diversity of personnel because its leaders understand that "classical" white supremacist policing requires drastic *reform* to remain politically and institutionally viable.

Thus, as the multiculturalist tenor of police recruitment becomes increasingly conspicuous through the LAPD's public relations efforts (via billboards, conferences, parades, and the Internet), its white supremacist substructure remains a primary determination of on-the-ground policing. The LA-based organization Youth Justice Coalition (YJC), formed and led by criminalized young people of color who are survivors of the local policing-criminal justice nexus, offers a clear rebuttal to the LAPD's veneer of diversity, sensitivity, and reform. According to YJC, whose formation has critically accompanied the multiculturalist "turn" of the LAPD's policing phenotype, the political and ideological structures of white supremacist criminalization are firmly ensconced in the biographical and geographic landscape of the city:

> YJC members, ages 7 to 24, are the young people L.A. has labeled as criminals, gangstas, thugs and hoodlums—in other words, we're basically considered trash. To most people, we are invisible and forgotten, locked away in dusty corners of LA County, behind barbed wire and concrete—in juvenile halls, county jails, camps and youth authorities. We've been pushed out of the school system into Continuation Schools and Probation Schools where the teachers are overworked and under-trained, the books and materials are in short supply, and there are more Probation Officers than guidance counselors. We report to Probation and Parole on the regular, and have gotten use to routine police searches and peeing in a cup on demand.[17]

Since the 1990s, the LAPD's apparent change of orientation in favor of a phenotypically diverse personnel has worked to facilitate, sustain, and enhance the systemic, everyday processes of racist criminalization: Los Angeles has long boasted the largest imprisoned population of any county in the United States, a carceral toll overwhelmingly composed of poor blacks and Latinos (including youth, undocumented migrants, and those jailed while awaiting trial).[18] Further, throughout the late-1990s, the LAPD Rampart Division was involved in multiple conspiracies to fabricate and steal evidence (including over $1 million of cocaine) and effectively framed hundreds of innocent people (more than 100 of whom have since been released from prison and jail after courts overturned their convictions). The Rampart Division was also engaged in numerous, sustained incidents of civilian shootings, in-station beatings, and street tortures. In fact, the "Rampart scandal" was notably marked by the *prevalence* of black and Latino officers as both central figures and implicated coconspirators in the investigation and criminal proceedings that revealed the depth and scope of this organized racist state violence.[19] More recently, the widely broadcast police violence of the 2007 May Day repression, in which LAPD riot control officers—many of them Latino—attacked immigrant rights activists and other civilians during a peaceful gathering at MacArthur Park,[20] further demonstrated that the institutional logic of racist state violence is not reducible to the uniformed, reactionary white cop.

Here, I would suggest that a more nuanced conception of the relation between the systemic logic of white supremacy—*which is not reducible to the white body itself*—and the strategic institutional reformism invested in transactions of racial bodies—in which the boundaries of public discourse (e.g., "diversity") are defined by the presence, marginality, and absence of particular bodies in certain institutional spaces—might help delink white supremacy's "substructure" (including its logics of violence and ordering) from reductive notions of the white supremacist "phenotype" (the white body). Put differently, I am asking us to consider how white supremacy is fully engaged in the organization of systemic violence and the social subjection of white civil society's historical racial antagonists, *even and especially* as its institutional forms—including its political and intellectual leadership—display a (relative) "diversity" of nonwhite bodies. What new political languages, organizing strategies, and conceptual frameworks can enable an adequately critical, ambitiously transformative response to this era of "multiculturalist white supremacy"?

REFORM

I have argued elsewhere[21] that insisting on a "definition" of white supremacy that is absolute, transhistorical, and hermeneutically sealed does not generate adequately dynamic critical activist or scholarly work: as a determination of different kinds of social formations, white supremacy's matrices of institutional mobilization, conceptual apparatuses of civilizational, national, and subjective ordering, and grammars of articulation (both as a rhetoric of power and common sense arrangement of everyday rule) too consistently flex and change in response to the specific political expediencies and social crises of different historical moments. Hence, the conceptual and political apparatus of an authentically antiracist, anticolonialist, and/or antiwhite supremacist praxis in one period—for example, the antiapartheid and antisegregationist work of "racial equality"—may be seized as a primary vehicle for white supremacist reaction and revivification in others.[22] In this sense, when the White Camelia Knight of the Ku Klux Klan (WCKKKK) rails against the alleged post-civil rights tyranny of "racial equality" in one polemical breath, it is also lamenting the *exclusion* of whites from the projected civil entitlements and privileges of this very same racial equality—it *desires* the perverse "racial equality" that it projects as the accumulated cultural capital of "nonwhites."

> A Double [*sic*] standard is when rules do not apply equally to all groups. It can also be a set of principles allowing greater opportunity or liberty to one group than to another.
>
> Today, Whites face a wide range of double standards. Whites are told that we should not think in terms of our race with pride, but non-whites are urged to promote their race. Whites are told that we should step aside in the political arena and allow non-whites more political power. At the same time non-whites are urged to get more involved and to take greater authority in politics. Non-white politicians openly say that in areas were [*sic*] their numbers are in the majority they should be the only ones that should be allowed to represent

their people, while at the same time whites are told when it comes to politics we should be colored [*sic*] blind.[23]

The WCKKKK's variation on white supremacist social thought is not without nuance and ideological layering. Here I invoke it not because it represents a significant, popular, or pervasive ideological prototyping of white supremacy, but rather because the WCKKKK's ideological resonances—and political compatibilities—with common, everyday enunciations of white supremacist social discourse—from the police precinct to the corporate boardroom—suggest a *continuity and coherence* to white supremacist sociality that requires substantive analysis and theorization. While the WCKKKK may be popularly disavowed for its most notorious pronouncements—"Racial suicide all in the name of equality is insane.... The Klan believes Whites are superior to the Non-Whites"[24]—the political structure of its allegedly "extremist" articulations is expansive, accommodating, and *common*. As I have argued elsewhere,[25] we can and should understand the KKK (and its multiple derivations) as part of the historical template for US nation building rather than its racist or white fundamentalist aberration.

Herein the theoretical and political conceptualization of white supremacy—the rigorous tracing of its discursive transformations, institutional possibilities, historical logics, *and changing embodiments*—requires a radical and critical attention, one that signifies the present tense capacities of white supremacy as a historical logic of social formation that is adaptable rather than rigid, "reformable" rather than narrowly reactionary.

White supremacy, as a social technology, involves *mobilizations of dominance* (most often, "racism") that reproduce and advance a mode of civilizational and national production: in fact, the making of these civilizational and national orders *are themselves often understood as "reforms,"* even if in historical retrospect we tend to think of those processes as profoundly reactionary or militaristically conservative. Manifest Destiny was a grandiose reformist ambition, intended to reform (and in places transform) the political trajectory, moral character, and racial stock of an emergent white nation-building project that stretched the landscape of a continent. The US pacification and colonialist modernization of the Philippines was another massive white supremacist reform, premised on a remaking of the native political, economic, and cultural order in a manner that further distended the white nation-building project such that it was to be gradually inhabited by the body of the civilizationally assimilated "Filipino."[26] Thus, rather than conceptually limiting white supremacy to an "ideological" artifact that is derivative of other social determinations, it is more useful to examine how it constitutes a social determination in and of itself, dynamically engaged with (rather than epiphenomenal to) the social determinations of the economic mode of production, heteronormative patriarchy, and so forth. It is precisely its reformist capacity that defines white supremacy's substructural "behaviors" in many historical moments.

The coherent *ideological* articulation of white supremacy is not the political or material prerequisite for the fatal productions of power that are the signature moments of its various mobilizations. Rather, the "superstructural" (e.g., institutional, military, ideological, governmental) articulations of white supremacy emerge from multiple, overlapping, and sometimes contradictory cultural and political *mediations* of the substructure: changing ensembles of institutional, intellectual/ideological, and cultural leadership (and insurgency) *arrange and coordinate* regimes of dominance that change, emerge, and become obsolete over time.

To consider white supremacy as a substructure, in this context, is to bring focal attention to the discursive logics that constitute its organic "mode of production," a process of racial sociality that persists across historical periods even as its outer expressions of social and political form—think segregation to "post-civil rights," racial colonialism to global neocolonialism—appear to mark genuine sociopolitical and cultural transformations. These logics are, of course, no less determining of the material social world simply because they compose a discursive, rather than economic, mode of production. To the contrary, following Hall's delicate rereading of Gramsci's social theory, when attempting to apply a notion like the "mode of production" to historically specific social formations,

> [T]he theorist is required to move from the level of "mode of production" to a lower, more concrete, level of application. This "move" requires not simply more detailed historical specification, but—as Marx himself argued—the application of new concepts and further levels of determination in addition to those pertaining to simple exploitative relations between capital and labour.[27]

Following Hall's lead vis-à-vis Gramsci, "racism" displaces "race" as a central term of engagement in this historical and analytical moment. Reiterating Gilmore's conceptualization of racism as "the state-sanctioned and/or extralegal production and exploitation of group-differentiated vulnerabilities to premature death,"[28] I should elaborate what I understand as the crucial feature of Gilmore's crystallized definition: racism, as the fundamental social technology of white supremacy, is a historically nuanced social practice that pivots on the institutionally multilayered, multiply embodied but always potentially genocidal production, reproduction, and "reform" of particular peoples' systemic *vulnerabilities* to biological, civil, and social death, collective historical trauma, and categorically manifested suffering. The mobilization of racism through the leverage points of fabricated and institutionalized vulnerabilities to violence, terror, and fatality definitively obviates (1) liberal academic notions of racism as a derivative of or synonym for intergroup "discrimination" and "prejudice," (2) empiricist social scientific renderings of racism as a necessarily *quantifiable* social harm, and (especially) (3) myopic and tautological formulations of racism as endemically or exclusively belonging to the white body, or as inclusively belonging to "racist"/"antiwhite" people of color.

Racism as systemic vulnerability implies more than an identifiable proximity to the institutional mechanisms of suffering and death. It also names a lived condition of alienation from the assumption of one's own bodily and spiritual integrity, a structure of immanent *bodily disintegration* that is as individuated and visceral as it is categorical and systemic. Finally, the fulcrum of racism as a dynamically formed violence and vulnerability disabuses the possibility of white supremacy's historical obsolescence or contemporary political marginality. By illuminating racism as a technology of dominance that is not endemically or exclusively coupled to a singular ("white") racial body or institutionality, Gilmore's working definition makes room for a radical retheorization of the very social logic and historical *telos* that has rendered racism as a determining force of the contemporary globality: put differently, white supremacy is as much a *reformist* as it is an overtly repressive logic of social organization and "change."

PHENOTYPE

While almost every variety of critical activist and scholarly social and political thought acknowledges that relations of dominance—from neoliberalism and imperialism to heterosexist patriarchy and bourgeois republicanism—do not necessarily manifest as overtly exclusionist, noncontradictory, and homogeneous (bodily) monopolies of power, this insistence on theoretical nuance has not sufficiently extended to conceptualizations of white supremacy. In fact, some of the most noteworthy and significant contemporary scholars of race seem to insist on replicating archetypal renditions of white supremacy that conflate it with past-tense social formations of "racial dictatorship" and literal white power monopoly.[29]

In opposition to this stubborn tendency toward theoretical and political minimization, I am arguing that white supremacy is a constitutive determination of *present tense* social formations, including and especially those that appear to have surpassed the cultural and political delimitations of classically recognized racist social arrangements like racial colonialism, apartheid, and de jure segregation. This present, no less, encompasses the *socially determining* powers that condense at the material surface of the white body and crystallize through the historical imaginary of the white social-historical subject.[30] Cedric Robinson's historicization of white supremacy's sociocultural roots in precolonialist Europe proves vital here. Robinson's groundbreaking book *Black Marxism* suggests that the political and theoretical animus of white supremacy, that which underwrites its coherence and consistency as a logic and *telos*, is inseparable from the incipient white European globality that in fact predates the epochal conquests of Africa and the Americas. He writes,

> The tendency of European civilization through capitalism was...not to homogenize but to differentiate—to exaggerate regional, subcultural, and dialectical differences into "racial" ones. As the Slavs became the natural slaves, the racially inferior stock for domination and exploitation during the early Middle Ages, as the Tartars came to occupy a similar position in the Italian cities of the late Middle Ages, so at the systemic interlocking of capitalism in the sixteenth century, the peoples of the Third World began to fill this expanding category of a civilization reproduced by capitalism.[31]

While the discrete elaborations of a *civilizational* white racial subjectivity arrived more expansively and overtly in the renaissance of European colonial conquest after the dawn of the seventeenth century, Robinson's landmark study lays bare how the *preconceptual* elements[32] of global white supremacy were catalyzed in and through Europe's own internal struggles away from feudal provincialism and toward capitalist civilization (and ultimately capitalist modernity).

Culminating in the pronouncements of the genocidal vision of Manifest Destiny[33] that marked the arrival of a qualitatively "Americanist" white supremacist global aspiration, it is precisely this endemically "racialist" structuring of the European civilizational project that precipitates the white supremacist *telos* that the hegemonic terms of "civilization" so consistently invoke. Thus, the accession of a conquest-bound white civilization as *the* compulsory paradigm on which modern civilizations would be calibrated was not singularly contingent on violent or exploitive encounters with indigenous non-European others, but was overdetermined at its origins by a logic of protoracial differentiation and the naturalization of oppressive civilizational dominance among Europeans. White supremacy, in this historical genealogy, is absolutely organic to the developmentalist *narratives* and coercively enforced global *procedures* of European capitalist modernity and civilizational progress: white supremacy is both the *sophistication* of the endemic European "racialism" of which Robinson writes, as well as the *marshalling of global communion* on which the (internally contested) emergence of a colonialist, conquest driven European/white global dominance was based.

To the extent that notions of "civilization" already require an imagination of the civilized and civilizing body—a working conception of its capacities, potentials, fashioning, and phenotype—it is crucial to distinguish the singularity of white supremacy as more than a benign or "inevitable" self-differentiation of Europeans from their others. As a uniquely *oppressive* and frequently genocidal imagination of the ascendancy of the white body, situated at the explosive historical convergence of emergent capitalism, the Enlightenment, rapidly accelerating global conquest, and complexly mobilized protomodern and modernist nationalisms, white supremacy is the articulation of a civilizational *condition* as much as a civilizational *destiny*.

It is the white body's biological and psychic fulfillment—formed in the discursive construction of its needs and desires *in the moment of civilization*—that frames the fundamental requirements of "civilization's" procedures: this is the condition that forms the premises of articulation for the most grandiose enunciations of the white body's historical *telos*, including and beyond its paradigmatic marking in Manifest Destiny. Further, this historical ensemble of civilizational procedures produces the white body's peculiar historicity as the module of a massive global fatality: the racial "body" of white supremacy is at once the signification, valorization, and material introduction of other bodies' structured susceptibility to extinction, disarticulation, and dispersal. Thus, to position white supremacy as fundamentally more than an ideological *consequence* of the racialist European civilizational project, but rather as its organic elaboration and, in the extended moment of this elaboration, the discursive matrix through which (white) "civilization" is rendered conceptually coherent and politically compelling, is to accent a particular theoretical consequence of Robinson's broader historical contention:

> What concerns us is that we understand that racialism and its permutations persisted, *rooted not in a particular era but in the civilization itself.* And though our era might seem a particularly fitting one for depositing the origins of racism, that judgment merely reflects how resistant the idea is to examination and how powerful and natural its specifications have become....As an enduring principle of European social order, the effects of racialism were bound to appear in the social expression of every strata of every European society no matter the structures upon which they were formed. *None was immune.*[34] [emphasis added]

White supremacy as a civilizational animus, its constitution of the historical *telos* of the white body's destiny and fulfillment, immediately suggests multiple premises of theoretical elaboration: first, the white body's emergence as the disciplinary prototype of the "human" (and thus as the model against which other peoples' "humanity" would be measured, ranked, or disputed) is not simply the imprint of the Enlightenment era's racial and protoracial articulations of rationality, civil subjectivity, and self-possession, but is the accumulated production of Europe's own internal violences and social transformations since at least the eleventh century—these violences and transformations are no less responsible for forging the disciplinary "white human" than are the traumas of racial conquest, colonialism, and slavery. Second, this production of the disciplinary white human is inseparable from the a priori sanctity (hence *singular* humanness) of the white body: the global institutionality and systemic enforcement of white existential entitlement to health, bodily integrity, and freedom of movement—as well as the white social body's relative *unfamiliarity* with and exceptionalist narrations of group-based premature death—is inseparable from this singular humanness. (That is, even the actuality of massive premature death among the white poor and socially dislocated is culturally and/or empirically incongruous with the systemic death of their black, brown, and indigenous "class" analogues.) Third, this historical genealogy facilitates a conception of *white life* as the durable, generally unspoken—and frequently unspeakable—framework of civil society as such.[35]

CONCLUSION: WHITE SUPREMACY'S ANNOUNCED (SELF-)OBSOLESCENCE

A critical scholarship of "the body" that apprehends the historical conditions of racism, white civil society, and the "racial state" must also reframe white supremacy's contemporary (postapartheid and postcolonial) modalities. I am suggesting here a further retheorization that can make sense of the changing global modalities of racism while paying close attention to changes and institutional transformations in the social modalities of global white supremacy. What is "white supremacy" in the aftermath of colonialism and apartheid, and how does it continue to constitute social formations that have formally altered or eliminated the central legal, governmental, military/policing, economic, and cultural systems that have conventionally institutionalized white supremacy as a relation of violence and dominance?

My argument for conceptualizing white supremacy as a sociodiscursive substructure is not only driven by an attention to the dynamic, shapeshifting historicity of the social forms it partly determines; it is also to suggest that white supremacy's political currency and effectiveness—its capacity to form the narrative-political terrain on which our notions of *historical agency* (including the "antiracist" and oppositional) form—is centrally enabled by a productive contradiction: white supremacy often *obsolesces* its terms of enunciation (e.g., the vernaculars of genocidal racial Manifest Destiny, biologized racism, segregation/apartheid) at the same time that its social logics of trauma, fatality, and disruption are *reproduced and enhanced*. Here we might extrapolate the contours of a white supremacist social-historical logic: the formal abolition and institutional disavowals of white racial and colonial dominance have been accompanied and succeeded by overwhelming material evidence that the *social ordering of racist subjection* transcends and survives—and may, in moments of political crisis, *necessitate*—the obsolescence of easily recognized, classical white supremacist social forms.

White supremacy is therefore not *reducible* to peculiar or singular physiological, social, and institutional phenotypes, to the extent that it is more usefully conceived as that which *animates* social formations and their shifts/transformations. To clarify, I am offering the notion of social and institutional "phenotype" here as a multilayered analytical rubric: I am concerned with the accumulations and dispersals of racially phenotyped bodies in sites of dominance *and* subjection (what do these sites *look like,* how is their particular racial geography generated in and through relations of dominance and subjection?), as well as with the dense and complex crystallizations of white supremacist sociality at the physiological site of the racial social subject. For example, how does white supremacy physiologically *produce and deform* particular bodies at the same time that it discursively categorizes and depicts them?

Further, I understand white supremacist sociality as the inscription of the white body's sanctity and abstracted white subject's transcendence and material entitlement within the conceptual matrices and mobilizations of "the social," including productions and imaginations of community, civil society, and governmentality.[36] Social and institutional phenotypes, in this sense, are the historical *production* and genealogical *outcome* of multiple determinations (warfare, political economy, ecology, etc.), but are enabled (i.e., materially prefaced and conditioned) and actively constituted by white supremacist sociality. Critical scholarly and activist work can, and in my view *must*, undertake an empirical and analytical inventory of how this sociality manifests in different political and geographic sites.

To reference prominent late-twentieth-century examples: while the shifts from "apartheid" to "postapartheid" and "segregationist" to "desegregated" social forms constitute a radical and undeniable change in the phenotypes of their respective social forms, these shifts do not automatically reflect—much less politically guarantee—a transformation in the social formation's constitutive logics of human subjection, a fundamental reconstitution of its ordering logic, or an altering of the social-historical substructure of racial subjection and white ascendancy. The white supremacist substructure, in other words, produces *multiple institutional phenotypes* and "works on" racial bodies in multiple ways: white supremacy's productions, deformations, solicitations, and mobilizations of racial bodies must not be conflated with the social ascendancy of "the white body."[37] There is, in other words, no *essential* "white supremacist body" within the institutional regimes of multiculturalist white supremacy—and to be clear, there absolutely *are* essential and prototypical white supremacist bodies within its "classical" renditions of conquest, colonization, racial slavery, and apartheid. Rather, it is the cultural spectacle and institutional coalescence of racial bodily diversity that has refurbished and innovated white supremacist social forms in the immediate (think "neocolonialism") and protracted ("post-civil rights," "postcolonial") aftermath of its white-monopolistic socialities.

My primary purpose here is not to quarrel on a piecemeal or point-by-point basis with differently premised explanations of continued black, indigenous, and formerly colonized peoples' enhanced modalities of suffering in the aftermath of classical white supremacist social forms like apartheid, racial slavery, genocidal conquest, or (settler) colonialism. That is, my analytical purpose is not to comprehensively address the insufficiencies of critical scholarly/activist analytics that might obscure, exclude, or theoretically minimize the social productions of an altered white supremacy in the historical present. Rather, I am schematically arguing for an *epistemological* departure that can strategically discomfort, disrupt, or productively displace existing critical analytics in those political moments when the social determinations of white supremacy appear to have been rendered marginal or obsolete by the accession of liberal, multicultural, and/or allegedly nonracial and desegregated institutional forms. What productive critical theoretical and scholarly activist praxis might emerge from a rigorously elaborated conception of white supremacy as a genealogical-political and social-phenotypic continuum rather than a discrete, distant historical periodization or obsolete institutional archetype?

NOTES

1. I am invoking here the allegedly historical and overtly ideological rubrics of the postracial, postracist, and post-civil rights United States as both popular cultural and public intellectual discourses. These discursive structures are animated by a racial nationalist political desire to transcend—hence permanently obscure, displace, or neglect—the nagging political antagonisms and social crises that remain inscribed by "race" as both a determination and mediation of institutionalized relations of dominance and human hierarchy. The primary arguments of this chapter suggest a fuller rebuttal of these "post-" rubrics.

2. Cedric Robinson, *Black Marxism: The Making of the Black Radical Tradition* (Chapel Hill: University of North Carolina Press, 2000).

3. See Stuart Hall, "Race, Articulation and Societies Structured in Dominance," *Sociological Theories: Race and Colonialism* (Paris: UNESCO, 1980), 305–345; Stuart Hall, "Gramsci's Relevance for the Study of Race and Ethnicity," *Journal of Communication Inquiry* 10, no. 2 (1986): 5–27.

4. Stuart Hall, Chas Critcher, Tony Jefferson, John Clarke, and Brian Roberts, *Policing the Crisis: "Mugging," the State and Law and Order* (London: Macmillan, 1978), 394.

5. Michel Foucault, *"Society Must Be Defended": Lectures at the Collège de France, 1975–76* (New York: Picador, 2003).

6. Ruth Wilson Gilmore, "Race and Globalization," in *Geographies of Global Change: Remapping the World*, ed. R. J. Johnston, Peter J. Taylor, and Michael J. Watts (Malden: Blackwell, 2002), 261.

7. See generally Denise Ferreira da Silva, *Toward a Global Idea of Race* (Minneapolis: University of Minnesota Press, 2007); Winthrop D. Jordan, *White over Black: American Attitudes toward the Negro, 1550–1812* (Chapel Hill: University of North Carolina Press, 1968); David Theo Goldberg, *Racist Culture: Philosophy and the Politics of Meaning* (Cambridge, MA: Blackwell, 1993); George M. Fredrickson, *The Black Image in the White Mind: The Debate on Afro-American Character and Destiny, 1817–1914* (New York: Harper and Row, 1971); Audrey Smedley, *Race in North America: Origin and Evolution of a Worldview* (Boulder: Westview Press, 2007).

8. The work of David Roediger and Steve Martinot offers some of the more engaged scholarly studies of how white supremacy composes a specific political and cultural formation that is not reducible to the processes of racialization and racial formation. See Roediger, *The Wages of Whiteness: Race and the Making of the American Working Class* (London: Verso, 2007); and Martinot, *The Rule of Racialization: Class, Identity, Governance* (Philadelphia: Temple University Press, 2003).

9. Michael Omi and Howard Winant's historical understanding of racial formation and the "racial state" reproduces precisely the conceptual (and political) errors that I am trying to correct here. Namely, that "white supremacy" composes a relatively compartmentalized *moment* in the historical life of racial formation in the United States, and thus "racism" is a kind of "racial project" (or mode of racialization) that is *not* organically linked to the changing historical continuities of white supremacy as a logic of social domination. See especially pp. 69–76 in Omi and Winant, *Racial Formation in the United States: From the 1960s to the 1990s*, 2nd ed. (New York: Routledge, 1994).

10. Jared Sexton, *Amalgamation Schemes: Antiblackness and the Critique of Multiracialism* (Minneapolis: University of Minnesota Press, 2008).

11. See generally, Saidiya V. Hartman, *Scenes of Subjection: Terror, Slavery, and Self-Making in Nineteenth-Century America* (New York: Oxford University Press, 1997); Dorothy Roberts, *Killing the Black Body: Race, Reproduction, and the Meaning of Liberty* (New York: Pantheon Books, 1997); Andrea Smith, *Conquest: Sexual Violence and American Indian Genocide* (Cambridge, MA: South End Press, 2005); João Costa Vargas, *Never Meant to Survive: Genocide and Utopias in Black Diaspora Communities* (New York: Rowman and Littlefield, 2008); Angela Y. Davis, *Women, Race & Class* (New York: Vintage Books, 1983).

12. See Omi and Winant, *Racial Formation in the United States.*
13. "Los Angeles Police Department Sponsors Gay Games," Gay Games VII: Chicago 2006 website, accessed December 2008 at http://www.gaygame-schicago.org/media/article.php?aid=123.
14. "10th Annual International Criminal Justice Diversity Symposium," Los Angeles Law Enforcement Gays and Lesbians (LEGAL) website, accessed December 2008 at http://www.losangeleslegal.org/pages/symposium/symposium1-cover.shtml.
15. William Bratton and George Kelling, "There Are No Cracks in the Broken Windows: Ideological Academics Are Trying to Undermine a Perfectly Good Idea," *National Review Online*, February 28, 2006, accessed January 2009 at http://www.nationalreview.com/comment/bratton_kelling200602281015.asp. For an overview of the contemporary history of "Zero Tolerance" and "Broken Windows" policing, see Christian Parenti, *Lockdown America: Police and Prisons in the Age of Crisis* (New York: Verso, 2000).
16. "Los Angeles Police Department partners with CareerBuilder.com to increase diversity awareness and drive applications," careerbuilder.com website, accessed January 2009 at http://www.careerbuilder.com/job-poster/enterprise/case-study.aspx?articleid=CEN_0004LAPD&cbRecursionCnt=1&cbsid=418f701ce0cb45dda7e082bd6258b64b-285950240-wi-6&ns_siteid=ns_us_g_los_angeles_police_di_.
17. "About Us," Youth Justice Coalition website, accessed December 2008 at http://youth4justice.org/blog/.
18. Amanda Petteruti and Nastassia Walsh, *Jailing Communities: The Impact of Jail Expansion and Effective Public Safety Strategies* (Washington, DC: Justice Policy Institute, 2008), see especially pp. 23 and 26.
19. "Frontline: L.A.P.D. Blues," Public Broadcasting Service, airdate May 15, 2001. For historical context, see Kristian Williams, *Our Enemies in Blue: Police and Power in America* (Cambridge, MA: South End Press, 2007).
20. Joel Rubin, "11 LAPD Officers Face Discipline in May Day Melee," *Los Angeles Times* (online edition), September 17, 2008, accessed December 2008 at http://articles.latimes.com/2008/sep/17/local/me-mayday17.
21. Dylan Rodríguez, "White Supremacy as Substructure: Toward a Genealogy of a Racial Animus, from 'Reconstruction' to 'Pacification,'" in *State of White Supremacy: Racism, Governance, and the United States*, eds. Moon-Kie Jung, João H. Costa Vargas, and Eduardo Bonilla-Silva (Palo Alto: Stanford University Press, 2011), 47–76.
22. By way of example, Manhattan Institute Scholar John McWhorter represents this political-intellectual tendency in his "Black neoconservative" texts *Losing the Race: Self-Sabotage in Black America* (New York: Free Press, 2000) and *Authentically Black: Essays for the Black Silent Majority* (New York: Gotham Books, 2003).
23. White Camelia Knight of the Ku Klux Klan, "Racial Equality," accessed September 2008 at http://www.wckkkk.org/eql.html.
24. Ibid.

25. See "Introduction: American Apocalypse," Dylan Rodríguez, *Forced Passages: Imprisoned Radical Intellectuals and the Prison Regime* (Minneapolis: University of Minnesota Press, 2006).

26. Rodríguez, "White Supremacy as Substructure"

27. Stuart Hall, "Gramsci's Relevance for the Study of Race and Ethnicity," in *Stuart Hall: Critical Dialogues in Cultural Studies*, eds. Kuan-Hsing Chen and David Morley (London: Routledge, 1996), 414.

28. Gilmore, "Race and Globalization," 261.

29. While there are myriad examples of this scholarly tendency, Omi and Winant's periodization in their paradigm-setting *Racial Formation in the United States* is worth critically revisiting precisely due to the wide influence of racial formation theory in the last fifteen or so years. On the one hand, the authors neatly compartmentalize what they call the period of US "racial dictatorship" between the years of 1607–1865, when "most non-whites were firmly eliminated from the sphere of politics" (65–66). On the other, they conceptualize "white supremacy" rather narrowly as the ideological premise of a fringe "Far Right" political reaction to the aftermath of the liberal reforms of the civil rights movement. In their view, beginning in the 1980s this Far Right attempted to *"revive"* white supremacy by "reassert[ing] white identity and reaffirm[ing] the nation as 'the white man's country'" (120). In my analysis, this historical and ideological flattening of white supremacy is both theoretically simplistic and politically dangerous, to the extent that Omi and Winant's presentation does not take seriously the multiple ways that white supremacist social logics (and ideologies) have constituted and/or deformed the institutional conceptions of "democracy," "freedom/liberty," "citizenship," and so on.

30. Silva, *Toward a Global Idea of Race*.

31. Robinson, *Black Marxism*, 26.

32. See Goldberg, *Racist Culture*.

33. See Reginald Horsman, *Race and Manifest Destiny: The Origins of American Racial Anglo-Saxonism* (Cambridge, MA: Harvard University Press, 1981).

34. Robinson, *Black Marxism*, 28.

35. See Frank B. Wilderson, III, "The Prison Slave as Hegemony's (Silent) Scandal," *Social Justice: A Journal of Crime, Conflict, and World Order* 30, no. 2 (2003): 18–27.

36. See Foucault, *"Society Must Be Defended."*

37. Jasbir Puar's examination of "white ascendancy" in her significant work *Terrorist Assemblages: Homonationalism in Queer Times* (Durham, NC: Duke University Press, 2007) is in direct resonance with my argument here.

CHAPTER 4

MATERIALIZING HOPE: RACIAL PHARMACEUTICALS, SUFFERING BODIES, AND BIOLOGICAL CITIZENSHIP

JONATHAN XAVIER INDA

ON JUNE 16, 2005, the Cardiovascular and Renal Drugs Advisory Committee of the United States Food and Drug Administration (FDA) held a daylong meeting to discuss the new drug application for BiDil.[1] The drug was being considered for approval to treat African Americans suffering from heart failure, a condition in which the heart is unable to pump sufficient blood to the body's other organs. Gathered together, in addition to the members of the committee, were FDA consultants, representatives of NitroMed (the maker of BiDil), and a number of other guests, from medical scientists and academics to heart failure sufferers and spokespeople for various African American and minority civil rights, professional, and political organizations.

Following a series of speakers from NitroMed, who generally addressed the scientific evidence pointing to BiDil's efficacy, the committee opened up the floor for public comments. One of the speakers was Debra Lee, a forty-eight-year-old African American woman with heart failure. She was there to tell her story. "In 1999," she stated, "I had a heart attack. There was blockage in my heart. A stent was inserted. In early 2003 I noticed a change in my health—coughing continuously; being visibly short of breath; walking short distances tired me out; waking up in the middle of the night; sleeping in a chair because I felt as if I would suffocate if I laid down."[2] Doctors tested

Lee for various conditions. And in August 2003, she was diagnosed with congestive heart failure.

Later that year, Lee was offered a chance to participate in the African American Heart Failure Trial (A-HeFT), a clinical trial cosponsored by NitroMed and the Association of Black Cardiologists (ABC) to test BiDil in self-identified black patients. She quickly said yes. How did Lee feel at the time of her testimony? "I feel fabulous," she narrated. "No more shortness of breath; I am able to walk and exercise without resting; I can sleep in my bed at night; I am working more hours at the Indianapolis Museum of Art; I have more energy."[3] And to what did she attribute this turnaround? "It is my strong faith in God," Lee said, "and a little pill called BiDil. I believe this pill is helping my heart to pump stronger.... In my opinion, this pill has changed so many things for me, given me a new lease on life.... I believe I have another 40 years or so to live my life to its fullest."[4]

The story that Lee narrated is at once about a suffering body and a body of hope. It is about the pain that comes with a debilitating, chronic, and fatal condition such as heart failure: the shortness of breath, the persistent coughing, the fatigue, the increased heart rate, and the list goes on. And it is about the expectation that BiDil will not only prolong life but also improve the quality of one's existence. This mobilization of suffering and hope was not unique to Lee's story. The imagery appeared time and again in other testimony delivered during the open public hearing. And in the end, it was key to the approval of BiDil. A number of speakers—most notably Dr. Shomarka Keita, an anthropologist affiliated with the Smithsonian Institution and the National Human Genome Center at Howard University—expressed concern that the case for BiDil seemed to be based on the idea that African Americans had a specific biological, and likely genetic, profile that made the drug more effective in their bodies than in the bodies of European Americans.[5] This implied that African Americans constituted a discrete biological/genetic grouping. Given the United States' historical experience with racial science, in which some groups were classified as biologically inferior and others as superior, to biologize African Americans and designate BiDil as a racial drug was to potentially open the door to stereotyping, discrimination, and marginalization. But relieving suffering and saving African American lives seemed to trump any concerns with the problems of biologizing race. The Cardiovascular and Renal Drugs Advisory Committee recommended the approval of BiDil for African Americans, citing among other things the "disproportionate burdens of heart failure in blacks" and the need to develop effective treatment in this population "in light of health disparities."[6] A week later, on June 23, the FDA, following the committee's recommendation, approved the drug "for the treatment of heart failure in self-identified black patients."[7] BiDil thus became the first drug endorsed by the FDA for a specific racial group.

In this chapter, I explore the role of African Americans in the making of BiDil. I suggest that while the use of race in the production of pharmaceuticals might have problematic elements, African American advocacy cannot be grasped as a return to "racial science": race is not invoked to set up hierarchies of difference nor to legitimate the subordination of blacks. Rather, the contemporary appeal of race in medicine for African Americans is better understood in terms of biological citizenship. Generally speaking, biological citizenship refers to the linking of rights and citizenship to matters of health, disease, and bodily suffering.[8] It thus includes any citizenship project in which ideas of citizenship are tied to beliefs about the corporeal, biological life of human beings. Such citizenship projects have become an important part of the political landscape in the United States, with individuals and communities increasingly and explicitly defining what it means to be a citizen in terms of their vital rights—their rights to life, health, and healing. African American support for BiDil is precisely about vital rights. It is about entitlement to health services, hope for better treatment, and helping suffering bodies. It is grounded in the belief that the African American community, historically excluded from the many benefits of biomedicine and disproportionately at risk for its harms, deserves access to life saving medications. Pivotal here is thus not the biologization of race, but the idea that medications targeted to African Americans are essential to materializing the hope of finding cures and achieving healthier bodies.

Intellectually, this project draws from and is meant to contribute to the growing body of interdisciplinary literature that has developed around the theme of vital or life politics. This work has emerged out of Michel Foucault's brief, but significant, writings and lectures on biopower.[9] In *The History of Sexuality*, Foucault remarks that biopower designates "what brought life and its mechanisms into the realm of explicit calculations and made knowledge-power an agent of transformation of human life."[10] Biopower thus amounts to nothing less than the taking charge of life by political power. It points to how political and other authorities have assigned themselves the duty of administering bodies and managing collective life. Influenced by Foucault's notion of biopower, scholars have mapped out a broad field of inquiry concerned with how the vital processes of human existence matter when it comes to politics. For them, what is often at stake in the management of individuals and populations is nothing other than life itself. Within this broad field, researchers have produced a number of important studies on a range of subjects, including health and disease, pregnancy and reproduction, humanitarianism, refugees and immigration, colonialism, the politics of death, genetics and genomics, and citizenship.[11] This chapter contributes to the analysis of the citizenship dimensions of life politics. It is concerned with the relation between vital rights and the bodily, biological life of the human.

A SKELETAL HISTORY OF BIDIL

Let me begin with a brief history:

BiDil is a fixed-dose combination of two generic drugs, hydralazine hydrochloride and isosorbide dinitrate.[12] Both generics are vasodilators that relax the smooth muscle in blood vessels and cause them to dilate. The end result is that blood is able to flow more easily through one's veins and arteries, and thus one's heart does not have to work as hard pumping. BiDil was developed as a therapy for heart failure, a progressive, chronic disease in which the heart's muscle gradually weakens and little by little loses its capacity to pump enough blood to meet the body's metabolic needs. This illness affects close to 5 million Americans, with an estimated 400,000 to 700,000 new cases diagnosed each year.[13] Of the total number, there is about an equal percentage of men and women, the majority are over 65 years old, and roughly 750,000 are African Americans.[14]

The origins of BiDil as a medication for heart failure can be traced to the 1970s.[15] In 1973, Jay Cohn, the University of Minnesota cardiology professor who created BiDil, hypothesized in *Circulation* that vasoconstriction—a narrowing of the blood vessels that restricts or slows the flow of blood—was an important factor in heart failure.[16] The reigning assumption at the time was that the disease resulted primarily from abnormalities in the heart's pump function and had little to do with peripheral circulation, or how well the blood moved through bodily vessels. Based on this hypothesis, Cohn and colleagues theorized that vasodilators might be a promising treatment for heart failure.[17] The idea was that by dilating the blood vessels and making it easier for blood to circulate, the heart would pump more efficiently, thus potentially allaying the symptoms and progression of the disease.[18] Vasodilators were put to the test in two US Department of Veterans Affairs sponsored studies led by Cohn: the Vasodilator Heart Failure Trials, or V-HeFT I and II. Neither trial, it should be noted, was constructed explicitly around race/ethnicity.[19] They each enrolled white and black patients (all men), but the goal was to test the efficacy of vasodilators on the general population, regardless of race. In V-HeFT I, which lasted from 1980 to 1985, 642 men with damaged cardiac function undergoing conventional therapy with digoxin and a diuretic were randomly assigned to three treatment groups. One group received a placebo, another the alpha-adrenergic blocker prazosin (a vasodilator), and the third the drugs hydralazine and isosorbide dinitrate (H/I).[20] While prazosin, like the placebo, failed to produce positive results,

V-HeFT researchers found that H/I could "have a favorable effect on left ventricular function and mortality."[21] The second V-HeFT trial took place between 1986 and 1991. It compared the effects of H/I with those of enalapril, an angiotensis-converting enzyme (ACE) inhibitor, in 804 men taking digoxin and diuretic therapy.[22] As was the case in A-HeFT I, results from this trial showed that H/I helped to reduce mortality from heart failure. However, this drug combination did not perform nearly as well as enalapril. In consequence, ACE inhibitors went on to become established as a first-line therapy for heart failure, while H/I came to be recommended for patients who did not tolerate or respond to these drugs.[23]

Significantly, in 1987, while still conducting A-HeFT II, Cohn applied for a patent to use hydralazine and isosorbide dinitrate as therapy for heart failure.[24] Issued in 1989, this patent granted Cohn a monopoly to market H/I as a heart failure medication through 2007.[25] Two years later, in 1991, he agreed to license the patent to Medco, an emerging North Carolina-based pharmaceutical company. Medco entered the agreement with Cohn based on the understanding that the FDA might accept the A-HeFT I and II results as the primary basis for the submission of a New Drug Application (NDA) for a combined H/I product.[26] The combined product would be called BiDil.[27] The NDA for BiDil was submitted to the FDA in 1996.[28] Shortly afterward, in February 1997, the FDA's Cardiovascular and Renal Drug Advisory Committee met to consider Medco's application. At the meeting, Cohn tried to convince the committee that there was a "strong basis for approval of BiDil for congestive heart failure. That is, that the combination of hydralazine and isosorbide dinitrate exhibits a survival benefit compared to placebo; that it exhibits a strong trend for improved exercise tolerance versus both placebo and versus Enalapril [and] that it produces a sustained improvement in ejection fraction."[29] The committee was not swayed, however. On a vote of 9 to 3, it rejected Medco's application for BiDil.[30] This is not to say the committee saw no clinical benefits to BiDil. Rather, the problem was that the V-HeFT studies did not generate the kind of statistical data necessary to meet the FDA's standards for new drug approval. Their numbers were deemed simply too muddled to be interpreted "with any degree of certainty."[31] Or, as Ralph D'Agostino, a statistician and committee consultant, puts it, the V-HeFT studies were "nice," but "there's so much going on that nothing really comes out clearly."[32]

Although the FDA's decision was a setback, this would not spell the end of BiDil. The drug would experience a rebirth as a racial medication. Following the FDA's rejection, Medco decided not to pursue the development of BiDil and handed over intellectual property rights back to Cohn.[33] At this juncture, Cohn, together with cardiologist Peter Carson and other colleagues, went back to reexamine the V-HeFT I and II data. Focusing on race, they found that hydralazine and isosorbide dinitrate appeared to work better on blacks than the ACE inhibitor enalapril. Specifically, they concluded the following:

> A retrospective analysis of data from V-HeFT I and V-HeFT II indicate significant baseline differences exist between black and white patients with heart failure, and race impacts importantly on the mortality reduction observed with vasodilator and ACE inhibitor therapy. The H-I combination appears to be particularly effective in prolonging survival in black patients and is as effective as enalapril in this subgroup. In contrast, enalapril shows its more favorable effect on survival, particularly in the white population.[34]

In light of these findings, Cohn approached NitroMed, a Boston area biotechnology company specializing in nitric oxide (NO) enhancing medicines, to take on development of BiDil.[35] Intrigued by the possibilities, NitroMed acquired the intellectual property rights to BiDil in 1999.[36] That same year, together with Cohn, the company began meeting with the FDA to explore developing the drug as a heart failure therapy for African Americans.[37] And a year later, Cohn and Carson applied for a new patent, whose rights they assigned to NitroMed, to use H/I in "treating and preventing mortality associated with heart failure in an African American patient."[38] This reformulated BiDil patent dealing specifically with blacks gave its holders exclusive marketing rights until 2020.[39] Following all this, in March 2001, NitroMed received a letter from the FDA stating that "[g]iven the subset finding and the overall trend toward a survival effect in VHeFT I, we believe a single, clearly positive study in a CHF population would be a basis for approval of BiDil® for the treatment of heart failure in blacks."[40] The stage was thus set for the launch of the A-HeFT.

Begun in May 2001, A-HeFT would be the first clinical trial focused strictly on African American men and women afflicted with heart failure.[41] This historic study was cosponsored by the ABC, who helped NitroMed recruit patients and medical researchers.[42] A-HeFT was designed to determine the safety and efficacy of BiDil on self-identified blacks. One thousand and fifty patients with New York Heart Association heart failure classifications III and IV (meaning moderate to severe) were randomized into two groups. One group received BiDil with a standard course of treatment, which generally included digoxin, diuretics, beta-blockers, and ACE inhibitors. The other was given a placebo in addition to the standard medication. A-HeFT was expected to last until 2005. However, its Data and Safety Monitoring Board, an independent group of experts assigned to monitor the trial, and Steering Committee halted the study in July 2004 due to the "significant survival benefit" of BiDil.[43] Results from A-HeFT indicated that, compared to the placebo group, individuals who took BiDil experienced a 43 percent decrease in the risk of mortality, a 33 percent reduction in the rate of first hospitalization resulting from heart failure, and a general improvement in their quality of life.[44] Shortly after the end of the trial, in December 2004, NitroMed completed the submission of a revised NDA for BiDil.[45] And in June 16, 2005, the Cardiovascular and Renal Drugs Advisory Committee of the FDA met to reconsider the drug. Backed by the A-HeFT data, and with the strong support of a number of African American organizations, including the NAACP and the Black Caucus, this time the panel strongly endorsed BiDil. With the FDA's official approval on June 23, the drug became the first racial pharmaceutical. Its package insert reads: "BiDil is indicated for the treatment of heart failure as an adjunct to standard therapy in self-identified black patients to improve survival, to prolong the time to hospitalization for heart failure, and to improve patient-reported functional status."[46]

THE BIOLOGIZATION OF RACE

Significantly, the story of BiDil has become embroiled in a bitter controversy over the meanings of race in genetics. A number of scholars, scientists, and lay people have expressed concern that the designation of BiDil as a racial drug biologizes African Americans.[47] The thinking here is that BiDil lends credence to the idea that race is somehow a biological category and that racial disparities in health are genetic in origin. At issue for these critics is the likelihood that striving for health equity through the biological lens of race will lead to the naturalization of health disparities, permitting biological explanations to overshadow social, economic, and ecological understandings of disease.[48] Moreover, they fear that medical and societal inequities will simply intensify with the advent of race-based medicine.

Given the history of racial science in the United States, from Tuskegee to sickle-cell anemia, this is no doubt cause for concern. I want to suggest, however, that although African Americans have certainly been biologized through BiDil, this biologization does not necessarily imply that races are discrete biological entities or that health disparities have a genetic foundation. In fact, most proponents of BiDil, including many African American supporters of the drug, seem to be very well aware that race is a rather imprecise marker for genetic differences among populations and that inequalities in health cannot be reduced to genetics. My main focus here will be with the role of African Americans in the biologization of race. But as a matter of contextualization, I also discuss NitroMed.

There is no doubt that NitroMed has been involved in the practice of biologizing and geneticizing African Americans. The company has repeatedly

made claims suggesting the reason African Americans suffer disproportionately from heart failure, perform poorly on ACE-inhibitors, and do well on
BiDil might be, at least partially, genetic in origin.[49] Such assertions turn
on a chemical compound found in the body called nitric oxide (NO). NO,
which forms in the endothelium (the cells lining the inner walls of blood
vessels), has been found to be a central factor in the dilation of the blood
vessels. NitroMed has suggested that African Americans may be more likely
than other groups to have an impaired ability to produce NO and thus to
maintain bodily vessels open. Ultimately, this dearth in nitric oxide production appears to predispose African Americans to developing heart failure. As
NitroMed puts it: "The high prevalence of deficient nitric oxide-mediated
vasodilation in black patients may explain in part why heart failure develops
disproportionately in black patients."[50] That BiDil, as opposed to other heart
failure medications, works well on African Americans has also been associated with nitric oxide. BiDil is believed to be an NO enhancer.[51] One of the
drug's components, isosorbide dinitrate, releases nitric oxide at the walls of
the blood vessels. However, its effect subsides rather quickly. Hydralazine,
BiDil's other element, is said to avert the loss of isosorbide dinitrate's effect.
This increased availability of nitric oxide resulting from BiDil is deemed
responsible for relieving the symptoms of heart failure in blacks.

Significantly, NitroMed has worked to link African Americans' nitric
oxide deficiency and the efficacy of BiDil to genetics. As part of A-HeFT,
NitroMed conducted a substudy called Genetic Risk Assessment in Heart
Failure (GRAHF). This study compared the frequency of genotypes related
to cardiovascular disease between A-HeFT participants and white heart failure subjects who took part in the University of Pittsburgh's Genetic Risk
Assessment of Cardiac Events (GRACE) trial. NitroMed presented its preliminary results as follows:

> Endothelial nitric oxide synthase (NOS3) gene researchers found that a
> majority of black patients in A-HeFT possess a specific gene variation that was
> observed in less than half of the white cohort from GRACE. NOS3, which
> encodes the nitric oxide synthesizing enzyme in the heart and vasculature,
> is important in hypertension and heart failure. The benefit of BiDil therapy
> on…mortality, first heart failure hospitalization and patient functional sta-
> tus…was seen in those possessing the specific NOS3 gene variation.[52]

The suggestion here is thus that BiDil's power to improve the condition of
heart failure sufferers is related to a genetic trait that might be more common in blacks than whites.

Following NitroMed's lead, African Americans have likewise been implicated in the process of geneticization. Particularly important in this process

has been the ABC. A number of medical researchers and clinicians affiliated with the organization have been involved in pointing to the role of genetics in heart failure and drug efficacy. Notable among these researchers are Clyde Yancy, currently medical director of the Baylor Heart and Vascular Institute in Dallas, and Keith Ferdinand, clinical professor in the cardiology division at Emory University. Through the ABC both were involved in A-HeFT and have served as advocates for BiDil. Of the two researchers, Yancy has probably been more associated with the geneticization of African Americans. He has repeatedly argued that "heart failure in Blacks is likely to be a different disease"[53] and that this difference is probably "physiologic"[54] in nature. By physiologic, Yancy essentially means genetic. As he puts it: "The physiology of the heart and blood vessels may differ, and there may be certain other differences that are determined by variation in genetic signals. Much work needs to be done in this area, but there is growing curiosity that a genetic basis may exist to explain some of the differences in outcomes that are seen in Black patients with heart failure."[55] Keith Ferdinand has similarly claimed that genetic variation might explain why standard heart failure therapies are less effective on blacks. In so doing, like NitroMed, he underscores the nitric oxide deficiency theory of black heart failure. Ferdinand notes, "Based on the pathophysiology of HF, blacks demonstrate less ability for endogenous nitric oxide to dilate peripheral blood vessels. This may be related to an increased frequency of genes that decrease the synthesis of nitric oxide in black persons.... Black patients, therefore, may have a decreased response to neurohormonal drugs such as ACE inhibitors, ARBs, and β blockers and an enhanced response to drugs that increase the delivery of nitric oxide."[56] While both Clancy and Ferdinand speak in tentative language, using words such as "likely" and "may," it is nevertheless clear that for them genetics has a role to play in black heart failure.

Notably, this geneticization or biologization of African Americans has been heavily critiqued by a number of social scientists and medical scholars. Their argument has at least two strands. One strand centers on the construction of race as a biological category. The contention here is that the existence of a drug such as BiDil, which is targeted to just African Americans, de facto legitimates the idea that races are biologically distinct entities—that blacks have particular, intrinsic biological attributes that function to differentiate them from other groups. For critics of BiDil, this conflation of race and genetics, given the history of racism in the United States, can potentially lead to a host of discriminatory practices. As Sharona Hoffman notes, "Public perception that scientific evidence has established that a particular 'race' is more vulnerable to life-threatening illnesses than others or does not respond to medications that cure others may reinforce negative race-based stereotypes and misconceptions. Particular populations may be seen as diseased or incurable, which could fuel the belief that there are inferior human subspecies."[57] The second strand of the argument deals with health disparities. The general claim is that focusing too much on genetics as a root of health disparities and emphasizing race-targeted pharmacology as a solution to such inequities unnecessarily takes attention away from the social causes of disease.[58] Ultimately, the concern is that such reframing of health disparities might lead to a misallocation of health care resources. Jonathan Khan, a prominent critic of BiDil, articulates this concern as follows:

> To the extent that there are real health disparities that correlate with racial groups, an over-emphasis on genetics as an explanation for the disparity can lead to a misallocation of intellectual and material resources. For example, hypertension (a primary cause of heart failure) is caused by a wide array of factors, some social and environmental, some genetic. There are disparities in the incidence of hypertension between blacks and whites. The drive to reduce such racial disparities to a function of genetic variation fuels a logic that would concentrate resources needed to redress disparities on pharmaceutical interventions that work at the molecular level, rather than addressing large issues of diet, behavior, racism, and economic inequity that also play significant roles in hypertension.[59]

According to BiDil's detractors, the practice of geneticizing African Americans can have rather significant and deleterious consequences.

The concerns raised about BiDil are certainly serious. There is no doubt the geneticization of African Americans can open the door to discriminatory treatment. However, it seems clear that neither NitroMed nor the medical researchers affiliated with the ABC advocate a crude genetic reductionism. For example, regarding health disparities, it is certainly the case that NitroMed suggested a genetic basis for heart failure among blacks, but the company has not ignored social and environmental causes. On their *Heart.Health.Heritage* website, responding to the question "Why are African Americans at greater risk?," NitroMed answers, "African Americans are at a much higher risk for heart failure in part because more of them develop high blood pressure and diabetes than other ethnic groups. Some scientists suspect that low levels of nitric oxide...may also play a role."[60] Furthermore, the company suggests that other likely causes of heart failure include poor access to good health care services, greater exposure to environmental pollutants, and a propensity to be overweight and exercise less. And regarding the construction of race as a biological category, it might be that NitroMed and ABC scientists argue blacks may be more genetically prone to developing heart failure, but this is not the same as saying that races exist as homogeneous, clearly differentiated genetic entities. Actually, the supporters of BiDil have generally emphasized that race is a social category. Clyde Yancy articulates this perspective most forcefully: "Race is neither physiologic nor scientific; rather it is a social construct that reflects a group of persons with shared ancestry and similar customs/lifestyles that also intermarry. Clearly, African Americans represent a heterogeneous group and there is no reason to believe that any single genetic trait is uniformly and exclusively distributed in African Americans.... However, within this group, it is conceivable that certain traits may be over represented and that these traits might contribute to disease."[61] NitroMed and the ABC also do not suggest that the BiDil will work just on African Americans, only that it appears to work better on/in this population.[62] Based on the GRAHF substudy, NitroMed in fact argues that the genetic markers that may predict a patient's response to BiDil "are not unique to any one racial group—although they may be more prevalent in one group versus another."[63] While not entirely unproblematic, the geneticization of African Americans here is not about absolute differences but about probabilities.

SUFFERING BODIES AND THE BIOCHEMICAL MATERIALIZATION OF HOPE

Although the biologization of race has been the most publicly discussed aspect of the politics of BiDil, for many African Americans, the citizenship dimensions of the drug are more significant. Over the past few decades, as Aihwa Ong has noted, health-based claims and matters of life have become central to the citizenship politics of the West.[64] In France, for example, ill health is now by and large deemed the most credible grounds for conferring legal recognition upon asylum seekers, with individuals suffering from life-threatening pathologies the most likely to gain official residency permits.[65] In Brazil, the state has followed "a policy of biotechnology for the people," universalizing access to life saving AIDS medication in the name of fostering the health of each and every individual.[66] And in the United States, individuals afflicted with a wide range of maladies—from AIDS and mental illness to chronic fatigue syndrome and muscular dystrophy—are taking action and being recognized on the basis of their damaged biology.[67] Across the West, individuals have come to make claims on and be recognized by political, medical, and other authorities in terms of "their 'vital' rights as citizens."[68]

African American advocacy for BiDil has to be understood in just such terms—that is, as a biological citizenship project. For many blacks, indeed, the drug has been less about genetics and more about achieving vital rights and relieving the pain and suffering of racial bodies. This is not to say that all African Americans have backed BiDil. There have certainly been many critics.[69] But black support for the drug has been rather robust. Organizations such as the Congressional Black Caucus, the National Association for the Advancement of Colored People (NAACP), the National Minority Health Foundation, the ABC, the International Society of Hypertension in Blacks, and the National Medical Association have all strongly advocated for BiDil.[70] What I aim to do here is focus on the suffering body and how African American hopes for recognition have been materialized, at least partially, in BiDil.

The biological suffering of African Americans has (at least) two dimensions. One dimension has to do with the understanding of blacks as being disproportionately afflicted—compared to other populations—with heart failure. Let me take as examples the ABC and the NAACP, the two most prominent black organizations supporting BiDil. Both organizations have stressed the taxing effects of heart failure on the African American community. For example, Keith Ferdinand, a cardiologist and Chief Science Officer of the ABC, states,

> Heart failure is a major health concern in the US and is particularly problematic in the African-American community where the disease has an earlier onset and exhibits increased mortality, even with hospitalization. Early onset is manifested primarily in middle-aged patients,...with a higher mortality than in whites. Between the ages of 45 and 64, African-American males have a 70% higher risk for heart failure than Caucasian males. African-American females between the ages of 45 and 54 have a 50% greater risk to develop heart failure than Caucasian females. It is estimated that there are approximately 700,000 African-Americans with heart failure and the number is expected to grow to 900,000 by 2010.[71]

The NAACP has made complementary claims. Juan Cofield, president of the New England Conference of the NAACP, has emphasized that "cardiovascular disease disproportionately affects African Americans, and access to [BiDil] will help to alleviate the burden of this disease."[72] Notably, the concern of the ABC and the NAACP has not been just with black suffering around heart failure. For them, this disease is one among others that disproportionately affects African Americans. Speaking at a 2007 press conference titled "The Healthcare Quality Crisis in America: A 21st Century Civil Rights Priority!," Hilary Shelton, director of the NCAAP Washington Bureau, asserts, "African Americans are 23% more likely to die from various types of cancer than Whites. African American and American Indian/ Alaskan Native infant mortality rates are more than 2 times higher than those for their Caucasian counterparts....African American diabetics are more than 3 times more likely than Caucasian diabetics to have a lower limb amputated."[73] Heart failure is thus located as part of broader pattern of African American biological suffering.

The second dimension of black suffering has to do with the question of biomedical neglect. One aspect of this neglect is the lack of available effective medications for treating African Americans with heart failure. The ABC and the NAACP have both emphasized how standard medications for heart failure do not appear to work well on African Americans. For example, the NAACP, in a resolution encouraging black patients with heart failure to ask their doctors about BiDil, notes that "some medicines approved for the treatment of heart failure, such as ACE inhibitors, appear to be less effective in Black patients."[74] And Malcolm Taylor, chair of the ABC Heart Failure Committee, states, "African-Americans may also be underserved in heart failure because some existing drugs, such as ACE-inhibitors and ARBs, approved for use in heart failure, may be less effective in African-American patients under certain conditions. These ethnic differences are documented in package inserts for enalapril (Vasotec), and for losartan (Cozaar), indicating less favorable outcomes in African American patients."[75]

Another aspect of biomedical neglect is the African American community's general lack of access to health care. The NAACP's Health Department, as a case in point, has highlighted how minorities are less likely to be insured than nonminorities, tend to receive substandard medical care, and often have limited access to mental health services.[76] And the ABC, in a report on cardiovascular health disparities coauthored with the National Institutes of Health, likewise highlights the inequalities in the health care system: "Structural factors and inequities in health care resources may hamper African Americans in gaining access to quality care. Large numbers of African Americans receive care in urban medical centers or community clinics. In both types of settings, resources may be lacking to deliver the high level of care required for good outcome."[77] Black suffering here is about both the unavailability of effective drugs and lack of access to health care.

This self-understanding of African Americans as a community of suffering has been highly significant. The belief that blacks suffer disproportionately from heart failure and that current therapies do not work well on them has underpinned calls for biomedical rights, namely access to BiDil. African American organizations have essentially promoted the drug as offering hope to a suffering population underserved by the medical/pharmaceutical establishment. That African Americans have invested BiDil with great hope was clearly in evidence during the FDA Cardiovascular and Renal Drugs Advisory Committee meeting. At that forum, representatives from several African American organizations spoke in favor of the drug. One point emphasized was the life-enhancing benefits of BiDil. Gary Puckering, for example, speaking on behalf of the National Minority Health Foundation, noted: "As evidenced by the A-HeFT results, approval of BiDil will have an immediate and positive impact on the health and quality of life of many

patients with heart failure....I support BiDil because it will extend the life
of many Americans with heart failure. I support it because it will improve
the quality of life of these patients."[78] Another point stressed was the potential for BiDil to reduce health disparities. In her remarks, Congresswoman
Donna Christensen notes, "Today, ladies and gentleman, you have before
you an unprecedented opportunity to significantly reduce one of the major
health disparities in the African American community and, in doing so,
to begin a process that will bring some degree of equity and justice to the
American health care system."[79] This espousal of BiDil ultimately culmi-
nated in rather forceful calls for the FDA committee to approve the drug.
As Lucille Perez, representing the National Medical Association put it:
"Given the disproportionate impact of cardiovascular disease on African
Americans, anything short of approval cannot be justified....The National
Medical Association urges this committee to recommend to the FDA that
BiDil be approved....African Americans continue to die...from heart
disease at the alarming rate of 78,000 a year. This number could be sig-
nificantly reduced if BiDil is brought to market as soon as possible."[80] For
some African American organizations, BiDil offers the black community an
opportunity to achieve a measure of health. The drug essentially functions
to materialize hope. It is seen as means to achieve vital rights for African
Americans and relieve their biological suffering.

A MODEST POLITICS OF LIFE

On January 15, 2008, NitroMed announced it was halting the marketing of
BiDil.[81] The company stated that while sales of BiDil had been increasing,
it did not have enough resources to mount the "marketing and sales effort"
necessary for the drug "to achieve its full potential."[82] The reality is that
from the outset NitroMed struggled with low sales of BiDil. Analysts had
estimated that revenue from the drug would total $130 million in 2006.[83]
But that year, BiDil only made $12.1 million.[84] And in 2007, sales of the
drug totaled a mere $15.3 million.[85] BiDil's market failure was a remarkable
development for a drug invested with so much promise.[86]

Despite BiDil's failure, the event of the drug has nevertheless been significant. Undoubtedly, the linking of race and pharmaceuticals, as critics of
BiDil suggest, has problematic elements. However, one should not dismiss
or ignore the biological citizenship politics of the drug. Whatever problems
there were with BiDil, the drug was unquestionably implicated in a politics
of life. It was a politics in which recognition from and demands on political and other authorities were based on the vital needs and suffering of the
black body. It was a politics wherein African American's biological status was
mobilized to gain access to rights. For a community historically neglected

by pharmaceutical companies and biomedical authorities, the recognition of their biological suffering was an important achievement. Through BiDil, African American heart failure patients were given hope that their suffering could be relieved. They were offered the prospect that their lives could be prolonged. Of course, BiDil does not address the social and economic problems that many see as the root of black biological suffering. Indeed, as Didier Fassin notes, contemporary recognitions of the suffering body tend to have the effect of reducing human existence to its mere "physical expression"[87] and of imposing "a legitimate order defining citizenship on purely physiopathological grounds."[88] But BiDil was never intended as a solution for all African Americans' problems. The drug's biocitizenship politics were actually more modest. African American supporters of BiDil were well aware that the drug was only a biological fix. As such, BiDil's promise was simply to help prolong life and relieve biological suffering.

NOTES

1. Center for Drug Evaluation and Research, US Food and Drug Administration, *Cardiovascular and Renal Drugs Advisory Committee*, vol. 2 (Washington, DC: US Department of Health and Human Services, 2005).
2. Ibid., 218.
3. Ibid., 219.
4. Ibid., 219–220.
5. Ibid., 221–224.
6. Steven E. Nissen, "Report from the Cardiovascular and Renal Drugs Advisory Committee: US Food and Drug Administration; July 15–16, 2005," *Circulation* 112 (2005): 2046. The vote for approval was 9–0, with 2 people recommending the BiDil be approved for the general population.
7. US Food and Drug Administration, "FDA Approves BiDil Heart Failure Drug for Black Patients," news release, June 23, 2005, http://www.fda.gov/bbs/topics/NEWS/2005/NEW01190.html.
8. Adriana Petryna, *Life Exposed: Biological Citizens after Chernobyl* (Princeton, NJ: Princeton University Press, 2003); Nikolas Rose and Carlos Novas, "Biological Citizenship," in *Global Assemblages: Technology, Politics, and Ethics as Anthropological Problems*, ed. Aihwa Ong and Stephen J. Collier (Malden, MA: Blackwell, 2004), 439–463.
9. Michel Foucault, *The History of Sexuality*, vol. 1, *An Introduction* (New York: Vintage Books, 1980); Foucault, "17 March 1979," in *"Society Must Be Defended": Lectures at the Collège de France, 1975–1976* (New York: Picador, 2003), 239–264.
10. Foucault, *History of Sexuality*, 143.
11. João Biehl, *Will to Live: AIDS Therapies and the Politics of Survival* (Princeton, NJ: Princeton University Press, 2007); Lorna Weir, *Pregnancy, Risk, and Biopolitics: On the Threshold of the Living Subject* (New York: Routledge,

2006); Peter Redfield, "Doctors, Borders, and Life in Crisis," *Cultural Anthropology* 20, no. 3 (2005): 328–361; Didier Fassin, "The Biopolitics of Otherness: Undocumented Foreigners and Racial Discrimination in French Public Debate," *Anthropology Today* 17, no. 1 (2001): 3–7; Ann Laura Stoler, *Race and the Education of Desire: Foucault's History of Sexuality and the Colonial Order of Things* (Durham, NC: Duke University Press, 1995); Giorgio Agamben, *Homo Sacer: Sovereign Power and Bare Life* (Stanford, CA: Stanford University Press, 1998); Paul Rabinow, *French DNA: Trouble in Purgatory* (Chicago, IL: University of Chicago Press, 1999); Nikolas Rose, *The Politics of Life Itself: Biomedicine, Power, and Subjectivity in the Twenty-First Century* (Princeton, NJ: Princeton University Press, 2007); Petryna, *Life Exposed*; Rose and Novas, "Biological Citizenship."

12. NitroMed, "BiDil®," accessed March 23, 2008, http://www.nitromed.com/bidil/bidil.asp.

13. Heart Failure Society of America, *Quick Fact & Questions about Heart Failure*, accessed July 3, 2008, http://www.hfsa.org/heart_failure_facts.asp.

14. Wayne Rosamond and others, "Heart Disease and Stroke Statistics—2007 Update: A Report from the American Heart Association Statistics Committee and Stroke Statistics Subcommittee," *Circulation* 115 (2007): e69–e171; NitroMed, *Heart Failure Backgrounder* (Lexington, MA: NitroMed, n.d.).

15. Jay N. Cohn, "A-HeFT: Old Dog, New Endothelial Tricks," *Current Issues in Cardiology* 32, no. 3 (2005): 366–368.

16. Jay N. Cohn, "Vasodilator Therapy for Heart Failure: The Influence of Impedance on Left Ventricular Performance," *Circulation* 48 (1973): 5–8.

17. N. H. Guiha and others, "Treatment of Refractory Heart Failure with Infusion of Nitroprusside," *New England Journal of Medicine* 291, no. 12 (1974): 587–592.

18. The standard treatment for heart failure at the time was a regimen of digoxin, a drug that strengthens the contractions of the heart muscle, and diuretics, which help to manage blood pressure. But these drugs did little to ameliorate the suffering of people with heart failure and improve their condition. David Rotman, "Race and Medicine," *Technology Review*, April 2005, 62.

19. Jonathan Kahn, "How a Drug Becomes 'Ethnic': Law, Commerce, and the Production of Racial Categories in Medicine," *Yale Journal of Health Policy, Law, and Ethics* 4, no. 1 (2004): 1–46.

20. Jay N. Cohn and others, "Effect of Vasodilator Therapy on Mortality in Chronic Congestive Heart Failure: Results of a Veterans Administration Cooperative Study," *New England Journal of Medicine* 314, no. 24 (1986): 1547–1552.

21. Ibid., 1547.

22. Jay N. Cohn and others, "A Comparison of Enalapril with Hydralazine-Isosorbide Dinitrate in the Treatment of Chronic Congestive Heart Failure," *New England Journal of Medicine* 325, no. 5 (1991): 303–310.

23. Kahn, "How a Drug Becomes."
24. Jay N. Cohn. Method of Reducing Mortality Associated with Congestive Heart Failure Using Hydralazine and Isosorbide Dinitrate. US Patent 4,868,179, filed April 22, 1987, and issued Sept. 19, 1989.
25. Pamela Sankar and Jonathan Kahn, "BiDil: Race Medicine or Race Marketing?," *Health Affairs* 24 (2005): 456.
26. Medco, *1991 10-K*, accessed July 24, 2008, LexisNexis Academic.
27. Medco, *1992 Annual Report to Stockholders*, accessed July 24, 2008, LexisNexis Academic.
28. PR Newswire, "Medco Research Files NDA for BiDil(R)," news release, July 3, 1996, LexisNexis Academic.
29. Center for Drug Evaluation and Research, US Food and Drug Administration, *Eightieth Meeting of the Cardiovascular and Renal Drugs Advisory Committee* (Washington, DC: US Department of Health and Human Services, 1997).
30. Center for Drug Evaluation and Research, US Food and Drug Administration, *Cardiovascular and Renal Drugs Advisory Committee 80th Meeting: Minutes* (Washington, DC: US Department of Health and Human Services, 1997).
31. Ibid., 2.
32. Reuters Health Medical News, "FDA Panel Votes against Medco CHF Combination Drug," news release, February 28, 1997, LexisNexis Academic.
33. Kahn, "How a Drug Becomes," 16–17.
34. Peter Carson, Susan Ziesche, Gary Johnson, and Jay N. Cohn, "Racial Differences in Response to Therapy for Heart Failure: Analysis of the Vasodilator-Heart Failure Trials," *Journal of Cardiac Failure* 5, no. 3 (1999): 182.
35. NitroMed, *BiDil® (Isosorbide Dinitrate and Hydralazine Hydrochloride) Tables. NDA 20-727. FDA Advisory Committee Briefing Document* (Lexington, MA: NitroMed, 2005), 16.
36. PR Newswire, "NitroMed Acquires BiDil(TM) New Drug Application for Treatment of Congestive Heart Failure," news release, September 10, 1999, PR Newswire.
37. NitroMed, *BiDil® Tables.*
38. Jay N. Cohn and John Carson. Methods of Treating and Preventing Congestive Heart Failure with Hydralazine Compounds and Isosorbide Dinitrate or Isosorbide Mononitrate. US Patent 6,465,463, filed Sept. 8, 2000, and issued Oct. 15, 2002.
39. Sankar and Kahn, "BiDil," 458.
40. Quoted in NitroMed, *BiDil® Tables,* 16–17.
41. NitroMed, *BiDil® Tables.*
42. NitroMed, *Following the Trail of Evidence: The BiDil® Story* (Lexington, MA: NitroMed, n.d.).
43. NitroMed, "BiDil® A-HeFT Trial," accessed March 23, 2008, http://www.nitromed.com/bidil/aheft.asp.

44. Anne L. Taylor and others, "Combination of Isosorbide Dinitrate and Hydralazine in Blacks with Heart Failure," *New England Journal of Medicine* 351, no. 2 (2004); 2049–2057.
45. NitroMed, *BiDil® Tables.*
46. NitroMed, *BiDil® Package Insert* (Lexington, MA: NitroMed, 2005).
47. Kahn, "How a Drug Becomes."
48. Duana Fullwiley, "The Molecularization of Race: Institutionalizing Human Difference in Pharmacogenetics Practice," *Sciences as Culture* 16, no. 1 (2007): 1–30.
49. NitroMed, *BiDil® Tables,* 14–15.
50. Ibid., 15.
51. NitroMed, "BiDil®."
52. NitroMed, "Genomic Analysis," accessed March 23, 2008, http://www.nitromed.com/bidil/genomic.asp.
53. Black Issues in Higher Education, "Heart Failure in Blacks May Signal a 'Different Disease,'" *Black Issues in Higher Education* 19, no. 9 (2002): 26.
54. Clyde Yancy, "The Role of Race in Heart Failure Therapy," *Current Cardiology Reports* 4, no. 3 (2002): 218.
55. Black Issues in Higher Education, "Heart Failure," 26.
56. Keith C. Ferdinand, "The Isosorbide-Hydralazine Story: Is There a Case for Race-Based Cardiovascular Medicine?," *Journal of Clinical Hypertension* 8, no. 3 (2006): 157.
57. Sharona Hoffman, "'Racially-Tailored' Medicine Unraveled" (Working Paper 05-32, Case Research Paper Series in Legal Studies, Case Western University School of Law, Cleveland, OH, 2005), 22.
58. Jonathan Kahn, "Getting the Numbers Right: Statistical Mischief and Racial Profiling in Heart Failure Research," *Perspectives in Biology and Medicine* 46, no. 4 (2003): 473–483.
59. Ibid., 479.
60. NitroMed, "Risk to African Americans," accessed March 24, 2008, http://www.hearthealthheritage.com/risk.asp.
61. Clyde Yancy, "The Association of Black Cardiologists Responds to 'Race in a Bottle': A Misguided Passion," *Scientific American*, July 30, 2007, http://www.sciam.com/article.cfm?id=the-association-of-black-cardiologists-responds-to-article&print=true.
62. Jay N. Cohn, "Discussion," *Journal of Cardiac Failure* 9, no. 5 (2003): S209; Ferdinand, "The Isosorbide-Hydralazine Story," 158.
63. NitroMed, "Common Questions," accessed February 28, 2008, http://www.bidil.com/pnt/questions.php.
64. Aihwa Ong, "Mutations in Citizenship," *Theory, Culture & Society* 23, nos. 2–3 (2006): 499–505.
65. Fassin, "The Biopolitics of Otherness."
66. Biehl, *Will to Live*, 8.
67. Rayna Rapp, Karen Sue Taussig, and Deborah Heath, "New Sites of Activism and Claims of Citizenship" (unpublished manuscript, Department of Anthropology, New York University, n.d.).

68. Rose and Novas, "Biological Citizenship," 441.

69. Notable African American critics include genetic epidemiologist Charles Rotimi, anthropologist Shomarka Keita, sociologist Troy Duster, and legal scholar Dorothy Roberts. An important target of their critique has been the insinuation that African Americans constitute a discrete biological/genetic population.

70. NitroMed, *A-HeFT Coalition Backgrounder* (Lexington, MA: NitroMed, n.d).

71. Keith C. Ferdinand, "Improving Care for African-American Heart Failure Patients," *US Cardiovascular Disease*, June 2006, 86.

72. Business Wire, "NAACP-New England Area Conference Commends Prescription Pathway Medicare Prescription Drug Plan on Availability of BiDil® for Plan Members," news release, January 22, 2007, http://findarticles.com/p/articles/mi_m0EIN/is_2007_Jan_22/ai_n17135663/print.

73. PR Newswire, "Civil Rights and Medical Leaders Call for Social Justice in Health Care," news release, September 27, 2007, LexisNexis Academic.

74. National Association for the Advancement of Colored People (NAACP), *Emergency Resolution No. 7: BiDil in Treatment Plan of Black Patients with Congestive Heart Failure* (Baltimore, MD: NAACP, 2005).

75. Malcolm P. Taylor, "Frequently Asked Questions about the A-HeFT Trial: Background for ABD Members," accessed March 26, 2008, http://www.abcardio.org/article_trial.html.

76. NAACP, "Health," accessed March 22, 2008, http://www.naacp.org/advocacy/health.

77. Association of Black Cardiologists and National Institutes of Health, *Eliminating Disparities in Cardiovascular Care and Outcomes: Roadmap to 2010* (Atlanta, GA and Bethesda, MD: Association of Black Cardiologists and National Institutes of Health, 2004), 10.

78. Center for Drug Evaluation and Research, *Cardiovascular and Renal Drugs Advisory Committee*, 211–212.

79. Ibid., 203–204.

80. Ibid., 258–259.

81. While NitroMed is discontinuing its sales and promotional activities for BiDil, the company is still keeping the drug on the market and available for patients. NitroMed is also in the process of developing an extended release version of BiDil, one that can be taken once a day instead of three times daily. NitroMed, "NitroMed Reports on BiDil XR(TM) Progress Following FDA Meeting, Announces Restructuring and Suspension of BiDil(R) Sales Force, and Retains Investment Bank to Advise on Strategic Options," news release, January 15, 2008, http://investors.nitromed.com/phoenix.zhtml?c=130535&p=irol-newsArticle&ID=1096799&highlight=.

82. Ibid.

83. Sylvia Pagan Westphal, "Heart Medication for Blacks Faces Uphill Battle," *Wall Street Journal*, October 16, 2006, http://www.post-gazette.com/pg/06289/730462-114.stm.

84. Ibid.

85. NitroMed, "NitroMed Reports Financial Results for Fourth Quarter and Full Year 2007," news release, February 14, 2008, http://phx.corporate-ir.net/phoenix.zhtml?c=130535&p=irol-newsArticle&ID=1108225&highlight=.

86. For explanations as to why BiDil failed see B. Séguin and others, "BiDil: Recontextualizing the Race Debate," *Pharmacogenomics Journal* 8, no. 3 (2005): 169–173.

87. Fassin, "Biopolitics of Otherness," 5.

88. Ibid., 7.

CHAPTER 5

Embodying Food Studies: Unpacking the Ways We Become What We Eat

Darcy A. Freedman

> Clearly, the eater's life and health are at stake whenever the decision is taken to incorporate, but so too are his [sic] place in the universe, his essence, his nature, in short his identity.[1]

No human is free from the need for food. Most of us consume food on a daily basis, in routinized processes that not only nourish and maintain our bodies but also produce and reproduce social worlds. Food is more than a compilation of vitamins and nutrients; it is physical and social, personal and political, and inanimate yet animating. In this chapter, I innovate an interdisciplinary, grounded, and embodied analysis to focus on food and food practices as vehicles for creating and recreating bodies that occupy different—and differently valued—positions within social hierarchies. I examine how we as eaters are agents in the production of socialized bodies that, in turn, influence both our biology and social organization. I conclude the chapter with a call for an embodied food studies, including research, practice, and policy related to the dynamic interplay among food, food practices, and health. This approach explicitly draws attention to the complicated and multidimensional relationship between food, bodies, and society.

WHAT SHOULD WE HAVE FOR DINNER?

The question "What should we have for dinner?" and the answers it inspires reflect and reproduce bodies and their social locations. Indeed, nations, faiths, races, classes, and genders are legitimized each time this question is answered.[2] "What should we have for dinner" is increasingly the focus of research conducted by health scientists examining factors that contribute to national and global "epidemics" of obesity. This line of research is often based on a one-dimensional understanding of food as nutrient and thus yields findings focused on behaviors and biological mechanisms to increase "good" and decrease "bad" nutrients entering our bodies. Resulting health promotion campaigns such as "eat five fruits and vegetables a day"[3] or "follow the food pyramid"[4] disembody eaters and the eaten. Yet, one eats cherries or Cheetos for simultaneous purposes, including health, sustenance, pleasure, pain, identity, companionship, convenience, history, culture, taste, cost, and so on. Eating is not simply a biological or behavioral experience, though it is fully biological *and* behavioral. Cherries and Cheetos enter our bodies through specific behaviors and their varied nutritional content has unique effects on our cells and organs.

Attempts have been made to develop a more capacious understanding of food-related decision making. In particular, several public intellectuals and "foodies" have refocused the lens to highlight the influence of agricultural policies and practices, globalization, industrialization, immigration, and capitalism on the question "what should we have for dinner?"[5] Bestselling author and journalist Michael Pollan began *The Omnivore's Dilemma* with this question, and he states that the goal of his book was to try "to figure out how such a simple question could ever have gotten so complicated."[6] For whom was the question "what should we have for dinner" ever simple to answer?

Answering this question illuminates the embodiment of food and food practices. We eat because of bodies that labored to plant, harvest, preserve, procure, and prepare foods for others. And these food practices both reflect and create social hierarchies. For instance, the American history of slavery, the very system that permitted whites to more freely determine what they would have for dinner, also served to constrain enslaved African descendents laboring in the fields without pay, without ownership, and without value. This history of domination restricted the foods African Americans could have for dinner and also detrimentally affected their health and wellbeing.[7] For a number of women, "what should we have for dinner" is a question whose answer may yield violence or abuse and also serves as a tool for perpetuating gendered inequities. For persons living in poverty, it is a recurrent reminder of one's location in the social hierarchy, drawing attention to varying degrees of insecurity—in food, shelter, income, safety, and other basic needs. Raced, gendered, and classed bodies are produced each time we answer this *not-so-simple* question.

The purpose of this chapter is to unpack the ways that we become what we eat through a grounded and embodied analysis of three case studies. My aim is to encourage a textual conversation, one in which readers question and requestion the active role we as eaters play in the creation and performance of Self and Other. The first case study is grounded in a middle-school cafeteria, a place wherein I as a teacher not only consumed many meals next to my seventh-grade students but also consumed my identity as a white person. The second is situated at a farmers' market; it allows for the examination of food as a tool for both reinforcing and resisting boundaries of race and gender. The final case study includes a scene from the HBO television series, *The Wire*. This scene focuses on a steakhouse-dining excursion, an event in which food was a tool for both bringing people together and setting them apart.

The case studies highlight the social relations of power manifested in and transmitted through food and food practices. They illuminate the ways that social constructions such as race, class, and gender are literally consumed and created through food, which in turn produce biologically raced, classed, and gendered bodies. My analysis involves an "interaction between the observer and observed."[8] I was an observer in all three of the instances, albeit in different roles: as a reflective teacher, researcher, and television viewer. Each interaction involved a range of participants (e.g., students, shoppers, and characters), all contributing to my understanding of the biosociology of food. The cases specifically focus on the ways that race, class, and gender are embodied through food and food practices.

CASE 1: HAMPTON MIDDLE SCHOOL CAFETERIA, BATON ROUGE, LOUISIANA

The clock struck 11:25 in the morning. This was our designated time to navigate to the school cafeteria about 100 feet away. This trek represented a time of transition; I was no longer the teacher at the head of the classroom. Instead, I became another participant in the daily dining rituals of twenty-three seventh graders. On many days, I walked in tandem with the students through the cafeteria lunch line and picked up the mint green five-compartment tray destined to be filled with the cooks' choice of the day. As a student many years prior, I rarely consumed food prepared by the school cafeteria. However, the cuisine at Hampton Middle School presented an aroma that was more like a Cajun restaurant than a school cafeteria. The inviting aromas of red beans and rice, jambalaya, turnip greens, and corn bread never ceased to tempt me to forgo my sack lunch and spend the $2.00 to purchase a meal prepared by the three African American cooks. As I passed through the lunch line, I smiled at the cooks, each wearing a crisp white uniform and a hairnet, and then picked up a carton of milk from the large metal cooler at the end of the line.

As I prepared to dine at the bleach-scented cafeteria table, I noticed my students' eyes navigating toward me. Their observations became even more intense as I removed the plastic coat covering the straw intended for my chocolate milk carton. I was unsure as to why they were so intrigued with me. Nevertheless, I continued to open the milk carton, place the straw inside the carton, and take a sip of chocolate milk. At this point, I could tell something was wrong. My students were rising in their seats as I swallowed the creamy brown liquid. Suddenly, one student blurted out, "Miss Freedman, you can't drink chocolate milk!" Clearly there was a problem with my beverage preference; however, I was unsure why chocolate milk was inappropriate, especially since about half of my students were drinking it. I placed my milk carton back into its special compartment on the green lunch tray and asked the student why it was not okay for me to consume chocolate milk. He stated confidently that they were afraid that I would turn black if I consumed chocolate milk. He continued to state quite emphatically that "White people drink white milk and black people drink chocolate milk." All of my students were African American and I was white. I could sense that they were indeed concerned that I had broken a food rule and harm was inevitable. I was bound to turn black.

This case exemplifies the compartmentalization of food into groups such as "food for us" and "food for them"; a concept identified over sixty years ago by Kurt Lewin, a pioneer in the fields of social psychology and action research.[9] In this case, the categorization of food was based primarily on race. Due to my status as a white person, chocolate milk was not a "food for me" and the consequence of transgressing this food rule was perceived to be dangerous. The security of my racial status was in jeopardy if I ingested chocolate milk. This brown beverage was not simply a source of calcium and other vital nutrients. It was also, and arguably even more so, the material manifestation of social status; and the embodiment of this liquid contradicted my identity. It was, to borrow Omi and Winant's term, racialized.[10] Racial meaning was being extended to a previously unclassified practice. Even more revealing was the double standard related to the chocolate milk/white milk food rule. Although it was disconcerting and even risky for a white woman to consume chocolate milk, the students were unfazed that many black students were drinking white milk. The risk of transgressing from white-to-black was different than from black-to-white.

The chocolate milk/white milk food rule revealed the ways that racial boundaries are created through food practices and also highlighted the hierarchical nature of socialized bodies. Transgression from black-to-white compared with white-to-black was less concerning for the students. It appeared that white race was valued; it was a social position that one ought to maintain. Or perhaps the students simply did not believe it was fully possible for one to transform from black-to-white. The students' responses highlight a complex understanding of race; one in which race is considered malleable rather than fixed, subject to (re)interpretation in a given social context.[11] One can perform, and in this process become, a new race through the practice of consumption. In contrast, one can resist racial categorization by refusing to ingest certain foodstuffs.

CASE 2: LINCOLN COURT BOYS AND GIRLS CLUB FARMERS' MARKET, NASHVILLE, TENNESSEE

In the community surrounding the Lincoln Court Boys and Girls Club, an individual is five times more likely to find tobacco products in local food stores than fresh tomatoes and three times more likely to find alcohol than bananas.[12] Communities such as this one are often described as "food deserts"—areas with limited or no access to stores selling healthy food products. In response to the low levels of food access in this community, a farmers' market was started using a participatory action research approach.[13] The following excerpt from my field notes is related to an interaction at one of the farmers' market.

The very last customers at the farmers' market were two staff from the Lincoln Court Boys and Girls Club. The director of the Club, an African American middle-aged man, purchased a large amount of fresh produce. While shopping he repeatedly told us that he is a "very good cook." He mentioned that he would be preparing a "vegetable feast" on Saturday with all of the produce purchased at the market. While the director was shopping another staff member, a white woman in her early twenties, came to the market. She was also buying a large quantity of vegetables. In addition to purchasing tomatoes, corn, squash, and bell peppers, the young woman purchased several pounds of turnip greens. At some point, the director noticed the woman's bags of turnip greens and began to laugh at her and state, "Girl, you don't know how to cook those [turnip greens]." The young woman responded by saying, "Mr. X, you've never had my cooking. My daddy loves my cooking."

This case highlights the biosociology of gender and race and implicates food and food practices as central to bodily and social formation. We both perform and become our gender and race through food. The case begins with the director of the Boys and Girls Club shopping at the farmers' market and repeatedly highlighting how his food practices diverge from gender norms. There are strong social norms related to men as meat eaters and to woman as food preparers.[14] This man was challenging traditional definitions of gender by emphasizing that he was going to prepare a "vegetable feast" with the foods from the farmers' market. He asserted his counteridentity by proclaiming: He is the cook and he is cooking vegetables. Although the director was comfortable challenging gender norms related to food, he was stringent about the ways that people perform race through food and food practices. In this case, turnip greens represented a marker of racial identity—of African American identity specifically. It was comical to this man that a young, white woman would consider purchasing and preparing turnip greens. She was not "doing race" appropriately.[15]

The young woman's response to the comment that she does not know how to cook turnip greens is also revealing. She is offended that her cooking skills are being questioned and reinforces her gendered identity as a food preparer by explicating that her "daddy" likes her cooking. She is young and unmarried; however, she is satisfactorily doing gender by being able to fulfill the appetite of the man in her life, her father.

The farmers' market provides a space through which discourse and action socially construct gender and race. This interaction reveals that gender is not merely a matter of X or Y chromosomes, but instead is performed and practiced and, in turn, transforms biology.[16] Likewise, race is far more than simply a biological trait. Food practices provide agency for human actors to gender and race their bodies.[17] This phenomenon is complex and intersectional involving—at the same time—both biological and sociological processes. Specific food items such as meat are not inherently gendered or raced. Yet these foods do provide different types of nutrients, which affect the body differently. Meat, for example, is high in protein, a macronutrient critical to muscle formation. And gendered norms related to food practices result in specific types of bodies gaining greater access to these nutrients.[18] The social valuation of meat results in men consuming more of it, which influences the size, structure, and composition of male and female bodies.

CASE STUDY 3: A SCENE FROM *THE WIRE*, BALTIMORE, MARYLAND

The final case study involves a scene from the fourth season of the acclaimed HBO television show, *The Wire,* a series focused on systems and structures that systematically limit opportunities for the urban poor.[19] This particular episode highlights the ways that taken-for-granted experiences such as dining in a restaurant are part of broader processes of creating difference. While all of the cases are limited insofar as they are representations of reality, this case is unique because its primary intent was for entertainment. The narrative provides a mediated window into "reality." The dining excursion is part of an episode entitled *Know Your Place.* This episode involves three African American students (Namond, Zenobia, and Darnell) with identified "conduct disorders" who earn a free dinner after successfully engaging in an experiment as a part of their alternative classroom. The winning students decide they want to eat steak for their reward so their mentor takes them to Ruth's Chris Steakhouse, a high-end restaurant specializing in beef products.

As the students make their way to the restaurant, Darnell declares: "I'm ordering the biggest quarter pounder. Fries too." Namond responds by saying, "This ain't Micky D's. You better be thinking t-bone steak, medium rare." Darnell responds by asking, "The blood all squirting out?" and Namond answers, "Nah, ain't no blood." After this exchange Darnell asks his mentor for some clarification on steak terminology, and the mentor states they will just have to ask the waiter. Namond responds to this comment by commenting, "Oh shit, there's a waiter."

As the students move into the restaurant they are greeted by a hostess and taken to their table. Along this path, the three youth and their mentor stand out against the backdrop of white diners. The rituals of upscale dining such as being welcomed by a hostess, having reservations, and listening to the menu specials combine to change the mood of the students from simple bewilderment to intense discomfort. The students are continually reminded of their out-of-placeness as they settle at their table. At one point, Namond laughs heartily and Darnell tells him, "Shhh…shut it down, other people will be looking at us." He does not want to draw additional attention to the group.

The waitress walks over to present the specials on the menu, including king salmon, sautéed free-range chicken, and wood oven roasted quail, among other things. With each new item, the students look increasingly uneasy and irritated. They seem to be second-guessing their reward.

The scene ends with the students departing from the restaurant with a look of defeat. Although Zenobia wanted to have her picture taken in front of the restaurant at the beginning of the night, by the end her desire has changed. She does not want to record this moment in history. The scene concludes with Darnell asking his mentor, "Yo, Mr. C, can we stop at McDonalds? That food wasn't right."

This case illuminates the role food and food practices play in shaping and securing one's location in the social world. In particular, it highlights the ways that social positionality is demarcated through food and food practices. Dining out at a restaurant provides a chance to reperform one's race and class since "each act of incorporation implies not only a risk but also a chance and a hope—of becoming more what one is, or what one would like to be."[20] However, this scene highlights that "passing" in this new position is often ephemeral. Food-related practices, ranging from how to dress to what to order to where to put one's napkin, were part of the cultural toolbox the students negotiated as they attempted to pass as proper guests at the restaurant. While they had some prior experience with a few of these practices, as time progressed they became increasingly less comfortable. In the end, they did not have the social capital to appropriately perform race and class in this social context.[21]

The students' experiences at the steakhouse exemplify the ways that not only racialized but classed bodies are created through food practices. French sociologist Pierre Bourdieu provides an analysis of class formation through various taste preferences. He states,

> Taste, a class culture turned into nature, that is, *embodied,* helps to shape the class body. It is an incorporated principle of classification which governs all forms of incorporation, choosing and modifying everything that the body ingests and digests and assimilates, physiologically and psychologically. It follows that the body is the most indisputable materialization of class taste, which it manifests in several ways.[22]

Taste, a product of one's class culture, was indeed in effect at the steakhouse influencing the students both physiologically and psychologically. While they incorporated foods representing a different class, in the end the students were not fulfilled by this new taste. "That food wasn't right" was Darnell's response to the steakhouse. He wanted to conclude the meal with a trip to McDonalds, a restaurant that was sure to satisfy his taste—a hamburger instead of an expensive cut of beef. In no uncertain terms, McDonald's was *his* place.

EMBODIED FOOD STUDIES

The case studies reveal, on the one hand, that drinking chocolate milk, shopping at a farmers' market, and eating steak at an upscale restaurant are actions that may quench physiological needs. These actions help us obtain requisite nutrients so we can develop, grow, reproduce, interact, and evolve. However, according to nutritional sociologist Claude Fischler, "food not only nourishes but also signifies."[23] The signifying aspects of the food practices introduced in this chapter illuminate the active role we as eaters play in the creation of Self and Other. In different ways and through different food practices, gender, race, and class were produced, affirmed, and consumed. Scenes from Hampton Middle School, the Lincoln Court Farmers' Market, and *The Wire* underscore the ways we *do* and *become* our biosociological selves through the production, ingestion, digestion, and assimilation of food.

The case studies reveal the continual back-and-forth relationship between social worlds and body formation as well as the multiple roles that food plays in this process. We are indeed what we eat, but we also eat what we are. An embodied food studies necessarily takes into account the complicated and multidimensional relationship between food, bodies, and society. Embodiment of food studies builds on the work of Nancy Krieger, a public health scholar who defines embodiment "as a concept referring to how we literally incorporate, biologically, the material and social world in which we live."[24] An embodied food studies is therefore fundamentally focused on the ways that food and food practices allow us to become biologically the social world in which we live. It emphasizes the ways in which our stomachs provide a space for the outside world to pass through us.[25] Building also on the work of medical anthropologist Clarence Gravlee, an embodied food studies explicates the ways that disparate experiences, social and physical environments, social structures and culture, global political economies, and ecologies influence food practices, which in turn become embodied in our genome, cells, and organs.[26]

An embodied food studies calls for health research that is focused on any aspect of food to necessarily include an analysis of the biosociology of food and food practices. Thus, a study focused on childhood obesity, for instance, would need to take into account the social context, conditions, and positions influencing food-related behaviors and processes as well as the biological responses to these dimensions. This would include the documentation of social factors as variables influencing food practices (Is there adequate access to food in the community?). It would also include an equally rigorous analysis of the cultural meanings and interpretations of food (What does drinking chocolate milk mean for one's identity?), the biosocial significance of food-related behaviors (How is social status maintained or challenged through the preparation of turnip greens?), and the risks associated with transgressing food rules (How does social positionality influence food purchasing patterns?). The analysis of food practices at Hampton Middle School, Lincoln Court Farmers' Market, and Ruth's Chris Steakhouse provide examples of how scholarship informed by an embodied food studies may address these domains of inquiry.

In addition, an embodied food studies draws attention to the concurrently structured pathways of embodiment described by Nancy Krieger. The first pathway is the "societal arrangements of power, property, and contingent patterns of production, consumption, and reproduction" and the second is the "constraints and possibilities of our biology, as shaped by our species' evolutionary history, our ecological context, and individual histories."[27] An embodied food studies emphasizes the social construction of power, property, and related patterns of production, consumption, and reproduction. It also examines the ways that these factors influence food and food practices as well as body shape, size, and structure. A food studies analysis focused on this pathway of embodiment may, for instance, explore the latent effects of slavery on contemporary views of agricultural products and messages focused on increasing consumption of fresh fruits and vegetables. Moreover, an embodied food studies examines the socially contingent constraints and opportunities of our biology. Scholarship informed by this pathway of embodiment may examine health outcomes such as diabetes or cancer through the lens of freedom.[28] If one lives in a context like Lincoln Court where you are more likely to find alcohol than apples and Cheetos than cherries, how does this constrain one's ability to make "healthy" choices, and how does this constraint literally "get under our skin" in the form of a host of diet-related health disparities?

The embodiment of food studies would add complexity to scholarship focused on the relationships among food, food practices, and health. Embodied food studies scholars observe this relationship through a teleidoscopic rather than a microscopic lens. Accordingly, embodied food studies scholars acknowledge that food and food practices are always in flux, continually changing as new lenses such as race, class, gender, nationality, region, religion, age, ethnicity, and their intersections are applied. Scholars informed by an embodied food studies would *not* assume that health messages such as "eat five a day" or "follow the food pyramid" are unbiased or neutral. Instead, any message pertaining to food would be attentive to its partiality by emphasizing its physical, cultural, psychological, and social dimensions and consequences. The subject of an embodied food studies is also fluid. Individual eaters and feeders remain key subjects, of course, but other subjects emerge as being equally important, including all sectors and actors in the food system (e.g., food producers, processors, distributors, marketers, and preparers), laws and policies related to food (e.g., taxes on food and school lunch policies), historical and contemporary practices of racial and class-based segregation, and gender norms and expectations.

In the face of an ever-increasing list of health disparities between social groups, particularly with respect to diet-related health conditions,[29] an embodied food studies offers new insights for scholars, practitioners, and policy makers. American culinary artist James Beard (1903–1985) once stated, "Food is our common ground, a universal experience." Scholars using an embodied food studies perspective would both agree and disagree with this assertion by suggesting that food is indeed a universal experience, but it is not "our common ground." In fact, food as a universal experience may be one of the most powerful tools for creating *uncommon* grounds, for establishing hierarchies in society, for producing and reproducing valued and devalued bodies, and for perpetuating health disparities among socially marginalized populations. Embodied food studies scholarship and praxis actively acknowledge the ways that food and food practices create uncommon grounds and provide a platform for dismantling the resulting hierarchies on "biological" and "social" levels and the connective tissue that binds them.

NOTES

1. Claude Fischler, "Food, Self and Identity," *Social Science Information* 27, no. 2 (1988): 281.

2. Caroline Walker Bynum, *Holy Feast and Holy Fast: The Religious Significance of Food to Medieval Women* (Berkeley: University of California Press, 1987); Wade Clark Roof, "Blood in the Barbeque: Food and Faith in the American South," in *God in the Details: American Religion in Popular Culture,* eds. Eric Michael Mazur and Kate McCarthy (New York: Routledge, 2001): 109–122; Daniel Sack, *Whitebread Protestants: Food and Religion in American Culture* (New York: Palgrave Macmillan, 2001); Jean Kilbourne, "Still Killing Us Softly: Advertising and the Obsession with Thinness," in *Feminist Perspectives on Eating Disorders,* eds. P Fallon, M. A. Katzman, and S. C. Wooley (New York: Guilford Press, 1994): 395–418; Hortense Powdermaker, "An Anthropological Approach to the Problem of Obesity," in *Food and Culture: A Reader,* eds. Carole Counihan and Penny Van Esterik (New York: Routledge, 1997): 203–210; Doris Witt, *Black Hunger: Food and the Politics of U.S. Identity* (New York: Oxford University Press, 1999).

3. Gloria J Stables, Amy F. Subar, Blossom H. Patterson, Kevin Dodd, Jerianne Heimendinger, Mary Ann S. Van Duyn, and Linda Nebeling, "Changes in Vegetable and Fruit Consumption and Awareness among US Adults: Results of the 1991 and 1997 5 a Day for Better Health Program Surveys," *Journal of the American Dietetic Association* 102, no. 6 (2002): 809–817.

4. Patricia Britten, Kristin Marcoe, Sedigheh Yamini, and Carole Davis, "Development of Food Intake Patterns for the Mypyramid Food Guidance System," *Journal of Nutrition Education and Behavior* 38, no. 6S (2006): S78–S92; Carole A. Davis, Patricia Britten, and Esther F. Myers, "Past, Present, and Future of the Food Guide Pyramid," *Journal of the American Dietetic Association* 101, no. 8 (2001): 881–885.

5. Frances Moore Lappé, Joseph Collins, and Peter Rosset, *World Hunger: Twelve Myths,* 2 ed. (New York: Grove Press, 1998); Marion Nestle, *Food Politics: How the Food Industry Influences Nutrition and Health* (Berkeley: University of California Press, 2002); Michael Pollan, *Omnivore's Dilemma: A Natural History of Four Meals* (New York: Penguin Press, 2006); Eric Schlosser, *Fast Food Nation: The Dark Side of the All-American Meal* (Boston: Houghton Mifflin, 2001).

6. Pollan, *Omnivore's Dilemma,* 1.

7. Ted A. Rathbun, "Health and Disease at a South Carolina Plantation: 1840–1870," *American Journal of Physical Anthropology* 74, no. 2 (2005): 239–253; Richard H. Steckel, "A Peculiar Population: The Nutrition,

Health, and Mortality of American Slaves from Childhood to Maturity," *Journal of Economic History* 46, no. 3 (1986): 721–741.

8. Kathy Charmaz, "Grounded Theory," in *Contemporary Field Research*, ed. R. M. Emerson (Long Grove, IL: Waveland Press, 2001), 337.

9. Kurt Lewin, *Resolving Social Conflicts: Field Theory in Social Science* (Washington: American Psychological Association, 1997), 289–300.

10. Michael Omi and Howard Winant, *Racial Formation in the United States from the 1960s to the 1990s*, 2 ed. (New York: Routledge, 1994).

11. William W. Dressler, Kathryn S. Oths, and Clarence C. Gravlee, "Race and Ethnicity in Public Health Research: Models to Explain Health Disparities," *Annual Review of Anthropology* 34 (2005): 231–252.

12. Darcy A. Freedman and Bethany A. Bell, "Access to Healthful Foods among an Urban Food Insecure Population: Perceptions versus Reality," *Journal of Urban Health* 86, no. 6 (2009): 825–838.

13. Darcy A. Freedman, "Politics of Food Access in Food Insecure Communities" (Dissertation, Vanderbilt University, 2008).

14. Jeffery Sobal, "Men, Meat, and Marriage: Models of Masculinity," *Food and Foodways* 13, no. 1 & 2 (2005): 135–158; Marjorie L. DeVault, *Feeding the Family: The Social Organization of Caring as Gendered Work* (Chicago: University of Chicago Press, 1997).

15. John L. Jackson Jr., *Harlem World: Doing Race and Class in Contemporary Black America* (Chicago: University of Chicago Press, 2001); Cornell West, *Race Matters* (New York: Vintage Books, 1993).

16. Anne Fausto-Sterling, *Sexing the Body: Gender Politics and the Construction of Sexuality* (New York: Basic Books, 2000).

17. Rachel Slocum, "Thinking Race through Corporeal Feminist Theory: Divisions and Intimacies at the Minneapolis Farmers' Market," *Social & Cultural Geography* 9, no. 8 (2008): 849–869.

18. See for instance, Amy Bentley, *Eating for Victory: Food Rationing and the Politics of Domesticity* (Chicago: University of Illinois Press, 1998).

19. Anmol Chaddha and William Julius Wilson, "Why We're Teaching 'the Wire' at Harvard," *Washington Post*, September 12, 2010.

20. Fischler, "Food, Self and Identity," 281–282.

21. Pierre Bourdieu and Jean Claude Passeron, *Reproduction in Education, Society and Culture*, trans. Richard Nice, 2 ed. (Thousand Oaks: Sage, 2000).

22. Pierre Bourdieu, *Distinction: A Social Critique of the Judgment of Taste*, trans. Richard Nice (Cambridge, MA: Harvard University Press, 1984), 190, emphasis in original.

23. Fischler, "Food, Self and Identity," 276.

24. Nancy Krieger, "Embodiment: A Conceptual Glossary for Epidemiology," *Journal of Epidemiology and Community Health* 59 (2005): 352.

25. Slocum, "Thinking Race through Corporeal Feminist Theory."

26. Clarence C. Gravlee, "How Race Becomes Biology: Embodiment of Social Inequality," *American Journal of Physical Anthropology* 139 (2009): 47–57.

27. Krieger, "Embodiment," 352.
28. Nancy J. Hirschmann, *The Subject of Liberty: Toward a Feminist Theory of Freedom* (Princeton, NJ: Princeton University Press, 2003).
29. Agency for Healthcare Research and Quality, "National Healthcare Disparities Report," ed. US Department of Health and Human Services (Rockville, MD: Agency for Healthcare Research and Quality, 2008).

CHAPTER 6

EPISTEMOLOGIES OF FATNESS: THE POLITICAL CONTOURS OF EMBODIMENT IN FAT STUDIES

KATHLEEN LEBESCO

> Body-building, pacemakers, artificial hearts, dialysis, cosmetic surgery, cinema, television, photography, airplane seat size, electric carts, wheelchairs, swim-suit material, sports equipment and other technological innovations have subtly altered the dimensions and markers of what counts as "natural" and "abject" bodies. Boundaries set by academic disciplines, research agendas,... and personal anticipations function as perimeter patrols for knowledge composition and embodied possibilities.
> —Carolyn DiPalma

WHEN I RECENTLY PROPOSED a new minor in Gender and Sexuality Studies at my small, urban liberal arts college, one crotchety colleague piped up quickly, tongue firmly planted in cheek: "Why not one in Vegetarian Studies too? So *everyone* can have their own course of study?" I was proposing a curriculum that, while not a traditional liberal art, is "old hat," relatively speaking—gender studies programs (in particular) have been fixtures at many institutions for decades now. I could only imagine what response might greet a curricular proposal based on my area of research expertise: Fat Studies. Surely my colleague would take it as this week's sign that the apocalypse is upon us. This interaction, haunted by the specter of debates over what counts as meaningful knowledge in the quest for institutional status and professional expertise, raises a number of fascinating issues that I address here on behalf of Fat Studies. First, what questions about the body guide knowledge in Fat Studies? How do bodies and specific types of embodiment relate to the emergence of Fat Studies? With what other "interdisciplines"—areas like Critical Race Studies, Disability Studies, Food Studies, Sexuality Studies, and Transgender Studies—is Fat Studies connected, and how? (How) are imaginings of the body and embodiment in these fields complementary or antagonistic? What about this "interdiscipline" invites fruitful linkages between activism and academia, and what are the constraints of such connections? Through an investigation of the history of scholarly interest in fatness and the nascent move toward a more organized and programmatic Fat Studies, my chapter charts the course of this exciting new area of embodied scholarship.

A PERSONAL HISTORY

I find myself in the fraught position of writing an essay on behalf of Fat Studies, despite having a complicated relationship to the topic and the field. I began doing research on social movements to reposition the fat person from subjection to subjectivity in about 1993, when I entered a PhD program in Communication at the University of Massachusetts, Amherst. There was no "Fat Studies" then to speak of, but I did manage to get my hands on a few important texts that shaped my thinking. Susie Orbach's 1978 self-help book *Fat Is a Feminist Issue* politicized the fat body in a way I had not encountered elsewhere, spinning the act of getting fat as a purposeful challenge to womanhood under patriarchy. (Unfortunately, it also presumed a causal link between compulsive eating and fatness and billed itself as an "astonishingly effective new approach to weight loss through satisfaction.")[1] *Such a Pretty Face: Being Fat in America* by sociologist Marcia Millman made me want to understand the characteristically intense reactions, particularly the moral outrage, to fatness that it described.[2] *Shadow on a Tightrope*, by Lisa Schoenfelder and Barb Wieser, which I found a little later, was saucy enough to use the "O" word (oppression) to describe the cultural workings on the fat individual, thus emboldening me to think deeply about the fat person as a political (and politicized) subject.[3] Dozens of more broadly concerned feminist tomes had convinced me that the body was a worthy site of study, a stage on which political forces intersected in personal form. Queer theory made me consider that struggles for subjectivity in an era of identity politics need not result in the assimilation of difference.[4] Armed with the convictions imparted to me by these texts, I began what has turned into a fifteen-plus-year scholarly affair with fat politics.

Nonetheless, when I first heard an enthusiastic colleague in Disability Studies use the term "Fat Studies" to describe what I was contributing to, it *did* faze me. What? I had always said I was a communication scholar with sympathies in sociology and political science. Feminist and queer theory were my guiding frameworks, and my *topic* was fatness. "Fat Studies" at first seemed overly, unnecessarily specific; when I studied a television program like *The Sopranos* I would not have described what I was doing as "*Sopranos* Studies," but rather "Media Studies" more broadly. Furthermore, I was interested in thinking about fatness in the context of other subject-marking physical and cultural experiences, and I thought that "Fat Studies" might be too restrictive a title for such intellectual pursuits. I hoped that the more inclusive term "Body Studies" would catch on; I liked how it enabled connections to and among disability, feminism, and queerness. I wimpily worried when I first heard that some respected colleagues were proposing a "Fat Studies" interest group at the Popular Culture Association/American Culture Association (PCA/ACA), imagining leaks to the public received by rolling eyes and easy dismissals of the intellectual and political work that we were doing in earnest. (Turnabout is fair play, after all: I had spent many years giggling at the caucus names on the PCA/ACA Calls for Papers—"Panics, Fads and Hostile Outbursts," "Cemeteries and Gravemarkers," "Circuses and Circus Culture," "Tarot in Culture," even "Claims for the Paranormal.") But now, several years later, I have simply gotten used to the moniker. I belong to a vibrant community of scholars and activists on a Fat Studies listserv. None of us were disciplinarily trained in Fat Studies, evidence of the absence of any institutionalized Fat Studies program. We are sociologists, political scientists, communication scholars, exercise physiologists, nutritionists, social workers, geographers, women's studies scholars, and more. Many, if not most, of us write books and articles on fatness and present work not only at the conferences of our larger disciplines, but also at the few formal spaces for work on fatness (like the Fat Studies group at PCA/ACA, a similar group at the National Women's Studies Association, and one at the American Sociological Association). Many of us even teach classes about fatness—such as "The Social Construction of Obesity" at UW Milwaukee's Department of Human Movement Sciences, "Fat and Society" in Anthropology at the University of Pennsylvania, "The Sociology of Body Size" at the University of California, Santa Barbara, and "Reframing Fat: An Introduction to Feminist Fat Theory" at York University in Toronto.

So when I say that I speak "on behalf" of Fat Studies, I mean that I am doing my best to represent a nascent, interdisciplinary field of scholarly inquiry that I would argue takes as its central mission not only to rethink the present day rhetoric of the "obesity epidemic" and to intervene in how our culture responds to fatness, but also to broadly reform thinking about the body and embodiment.

My personal positioning might be thought to further complicate my relationship to Fat Studies, but in fact, I do not think that it does. Sometimes I am fat, sometimes I am not. The work that I do—arguing for subjectivity for people regardless of their body size—means that my own experience of gaining or losing weight should not matter. Sometimes I cringe at being an effective front person for Fat Studies when I am of average weight, remembering the compromises in how young, attractive heterosexual white feminists were the only mildly acceptable proponents of women's lib. (Butch lesbians, women of color, grannies and trannies, thank you for applying, but your services are not needed at this time.) At other times, when I am fat, I find frustrating the easy dismissal of my interest in Fat Studies as merely self-reflexive and justificatory—an oft-repeated condemnation lobbed at interdisciplines that focus on bodies, subjectivities and identities by critics whose ideals of neutral, universal, unsituated scholarship seem naïve at best. Generally speaking, regardless of my own body size or shape, my commitment to the restorative subjectivity-building project of Fat Studies remains profound.

EMBODIMENT AND FAT STUDIES

Good feminist, I find myself annoyed by my own attempts to write about "the body"—*any* version of "the body," whether fat or otherwise. I conjure up phrases like "living in a fat body," then backspace over them in the blink of an eye. One lives in an apartment, a house, not a body; instead, one *is* a body. "Fat embodiment" seems to better get at what I am thinking about, but this is not because I desire to erase the fat body as both a subject and an object of knowledge. "The body" implies a universal body. I grow tired of retyping "body" again and again and try my word processing software's built-in thesaurus. I am not looking for synonyms related to "organization, bulk, or quantity" (at least not the way *they* mean bulk). So the closest is this unfortunate string: "corpse, dead body, cadaver, remains, carcass, stiff, deceased." I am not about to do a "search and replace" with any of these, but I am contented that my annoyance with "the body" is appropriate, alienated and unlively as its connotations are.

Philosophers Gail Weiss and Honi Fern Haber recognize a shift from thinking about "the body" as a "nongendered, prediscursive phenomenon that plays a central role in perception, cognition, action, and nature" to "embodiment" as "a *way* of living or inhabiting the world through one's acculturated body."[5] This framing is very useful to my imaginings of Fat

Studies. Weiss and Haber also claim that "the body is increasingly being identified as central to our sense of agency as well as a distinctive cultural artifact in its own right."[6] This helps me to avoid throwing the body out with the bathwater, appreciating the tentacles of Fat Studies that explore the "distinctive cultural artifact" of the fat body as well as those that examine the layerings of meaning that constitute fat embodiment. We need to examine both ends of the spectrum (and all stops in between), malcontent to retain only "the fat body" as our object of scrutiny, because

> by abjecting the "fat" body from the culturally constructed aesthetic domain, people and not just body parts are designated as the abject other, doomed to exist in those uninhabitable, unlivable regions that Butler reminds us are, in point of fact, densely populated. Indeed, these regions are not just inhabited by those who are considered or consider themselves to be overweight. In fact, I would argue, they are currently in danger of being overpopulated insofar as none of us can forever live up to what Audre Lorde calls "the mythical norm."[7]

Scholars of education Stephanie Springgay and Debra Freedman posit a productive relationship between body knowledge and more sophisticated understandings of individual and collective agency that provides a useful analog to Fat Studies. Springgay and Freedman argue that a bodied curriculum—one that "attends to the relational, social, and ethical implications of being-with other bodies differently and to the different knowledges such bodily encounters produce"—is disruptive and risky because of its fluidity and openness to uncertainty.[8] "A bodied curriculum," they argue, "not only resists the very notion of standards, hegemonic power positions, and categories of sameness, it dislodges and destabilizes 'the center' from which binaries and dualistic logic are produced and maintained."[9] These ideas are familiar to most who work in Fat Studies, where "the body" is never just a material reality but also the site of contested discourses about power, health, beauty, nature, race, class, and a bevy of other possibilities. As Joe Moran points out, "the body is both a material, biological entity and a cultural product which we can change."[10]

So what does the body have to do with Fat Studies? Both a lot and a little, as it turns out. Fat Studies concerns itself with the actual lived experience of fatness, as much as it attends to fatness as an economic, historical, political, social, and cultural construct. Noted fat activist (and founder of the Fat Studies listserv) Marilyn Wann sees the field as less about the body, per se, than an intervention in cultural reactions to the existence of fat people.[11] Some of the work in the field, like my own, is more theoretical in its treatment of the fat body, the thin body, and everything in between, which is likely a trait of my home discipline—I find that analyzing rhetoric about fatness is a better use of my talents than providing objective analyses of "facts" about fatness, particularly in light of my conviction that all knowledge is contextual. Other scholars in the humanities adopt a similar approach: amid the current panic about the "obesity epidemic," historian Elena Levy-Navarro holds that a fat history that questions seemingly transhistorical and natural bodily categories would also examine "the types of emotional, aesthetic, and political attachments we take for granted and, in so doing, help us foster very different types of commitments, attachments, and identifications."[12] By writing such a history, she hopes to render visible the questionable "objectivity" of our "facts" about bodies and bodily categories, noting that the early modern period is particularly useful for helping us to see how moralizing our current discourses of pathology are.

Other work by scholars hailing from fields such as exercise science, physiology, and nutrition is far more empirical, usually integrating data or descriptions of the form and function of fat bodies as purported truths/facts. Exercise physiologist Glenn Gaesser makes empirical arguments that question the supposed unhealthiness of fat. Gaesser says that we are wrong to focus on weight and dieting and should instead pursue metabolic fitness apart from weight loss. His work hopes to discredit "the myths that obesity is a 'killer disease,' that weight loss is good, that thinner is necessarily healthier, and that the height-weight tables measure something meaningful" with boatloads of contrary data.[13] In rescuing fat bodies from the dustbins of ill health, this approach imbues them with value. Such contested constructions advance alternative truth claims about fatness and fat bodies.

Likewise, Gina Kolata, a science reporter for the *New York Times*, has written a book considered important on the empirical side of fat studies, though she probably would not identify herself as a proponent of the field. Kolata uses words like obesity and overweight without much reflection, but at the same time she argues for a paradigm shift about fatness: "When health data have not supported the alarmist cries of a medical disaster in the making, could society perhaps let up on the beleaguered fat people?"[14] All in all, Kolata seems to advocate leaving fat people alone—perhaps even suggesting that increased weights might be a result of good health, rather than the opposite—and thus serves well the political project of Fat Studies (making space for fat bodies and room for fat subjectivity).

And then there is a middle ground, with scholars like political scientist Eric Oliver, sociologists Abigail Saguy and Kevin Riley, physical education professors Michael Gard and Jan Wright, and law professor Paul Campos doing both theoretical work and making empirical rejoinders to what most Fat Studies scholars understand as the world's faulty claims about fatness. More specifically, Oliver, Saguy and Riley, Campos, and Gard and Wright review and critique expert claims from the scientific and medical communities about "the obesity epidemic" as part of a larger project of urging readers to rethink their assumptions about weight and health.[15] They examine and question methodologies and assumptions at the microlevel while also attempting to ideologically rescue the status of fatness. This integrated approach to fat studies spills over into the activist realm, as well. Perennial scholars' favorite Marilyn Wann says, "In my life, I see myself attempting—by all sorts of means—to create and expand the livable space in which I can be a person (in my particular embodiment). It's both a physical and intellectual endeavor. The adversarial pairing is never (in my imagination) me vs. my body; it's always me (embodiment and all) vs. the ways of thinking that would keep me down."[16] Wann's comments speak clearly to the ways in which embodied scholarship in Fat Studies is both about the body and about larger cultural anxieties.

Communication scholar Michael Hecht's work on "layering perspective" is helpful in mapping out the landscape of Fat Studies, given these divergent approaches. Hecht advocates understanding social reality via many experiential layers ("such as cognitive or emotive; or based on behavior or spirituality; or reality as an 'objective' experience or influenced by social rules") rather than adopting one unified view from above.[17] Such a use of multiple perspectives at play in Fat Studies—theoretical, practical, empirical, and so on—allows for a deeper understanding of fat embodiment. Recognizing the shared political project, there is a lesser degree of infighting about methodology or epistemology across the wide variety of approaches that make up this field than in many other fields (and here I am thinking of my home discipline of communication, where interpretivists and objectivists are often at one another's throats). Applying Cheryl Nicholas's Hecht-influenced work on GLBTQ identity to Fat Studies, the following quote is apt: "Across ontologies, epistemologies and methodologies are the different disciplines that use their own 'lenses' through which to view [fat] identity. The layering process allows explanations offered by each discipline to be juxtaposed with each other."[18] Like intersectionality, the layering perspective provides a more nuanced perspective on the topic at hand.

Fat Studies offers these different perspectives on fat embodiment at the same time that it speaks to broader issues, including but not limited to theories about the body. Contributors to the Fat Studies moderated listserv argue that the field is about power and difference, naming the body as only one site, a site in which we subscribers happen to share a common interest. Like other fields of study—environmental studies, global economics, popular culture—the body is engaged, to be sure, but not necessarily the constant center of focus.[19] In the same way that Freud's cigar was perhaps, sometimes, just a cigar, the fat body might be understood as sometimes just a fat body—but usually its experience in the world is an intriguing indicator of our perspectives on other big topics, and it is hard to imagine a body having any meaning whatsoever divorced from its context. As Thomas Csordas argues, "If we are not studying the body per se, neither are we studying embodiment, but studying culture and self in terms of embodiment, just as we can study culture and self in terms of textuality. Thus, to work in a 'paradigm of embodiment' (Csordas 1990) is not to study anything new or different, but to address familiar topics—healing, emotion, gender, or power—from a different standpoint."[20] Csordas sees the most significant advances in the study of embodiment in cultural studies of health and illness—places where "bodiliness is most overtly problematized" with broad cultural relevance.[21] These works, like many in Fat Studies, have a keen eye for the body as ground on and through which culture exists; their arguments "are rarely limited to disease per se but also teach us about broader issues of self, emotion, religion, meaning, transformation, social interaction, institutional control of experience, and the human interface with technology."[22] Key works in Fat Studies also get at these broader tropes of human experience—and thus rather than diluting or splitting knowledge, as some conservative critics have charged of area studies, they expand and strengthen it.

So what is *not* fat studies? How does one delineate the field? Work about fatness that fails to recognize or situate fat people as subjects would be excluded; the corollary in women's studies is that nonfeminist work focused on women would not be included. Carolyn DiPalma asks how maintaining the body as a subject rather than an object of perception transforms the possibility of political inquiry—and this distinction pertains well to Fat Studies.[23] Fat Studies involves not simply taking fat and fat people as an object of study, but instead seeing the fat body as characterized by subjectivity, which is itself a political project, as well as broader sociocultural meanings surrounding body size and shape. "Embodiment is entirely misunderstood if it is cast as no more than a passive resultant of several more or less causal forces. Central to the idea of 'embodiment' is the notion of *agency*."[24] For this reason, Fat Studies scholars read as misguided most attempts to study fatness that do not presume agency on the part of fat people as an integrated mind-body unit. Fat Studies tends not to imagine the fat body as possessing any inherently disruptive logic or agency apart from its situatedness as a facet of fat personhood. (This is not to say that this is impossible for other scholarship, but it is not currently a feature of the Fat Studies landscape.) There is plenty of work in the sciences on "overweight" and "obese" bodies, but it fails the litmus test of subjectivity and thus it does not fall under the rubric of fat studies. Work that examines cultural meanings of "overweight" and "obesity," while remaining sanguine about the presence of fat subjectivity, is welcomed. In contrast, a growing corpus of literature in the humanities and social sciences (particularly notable for its authorship by feminist scholars) takes up the "problem of obesity" as one in which people are victimized by larger social, economic, and political factors and thus become fat/Fat;[25] again, although it deals with fat bodies, this alone does not characterize this work as hailing from Fat Studies because of the obliteration of agency it suggests. Whether framing fat people as depressed and depressing cogs in the capitalist machinery, the unwitting dupes of Big Food, or the unfortunate victims of an "obesogenic" environment, this work may appear to be Fat Studies from without. But from within, Fat Studies scholars are no more likely to embrace it as "what we do" than Women's Studies scholars are to cuddle up to Descartes (whose gendering of the mind-body split left women with the short straw) or queer theorists are to adopt as a mascot Martha Nussbaum (who derides them as preoccupied with frivolity instead of working toward real social change).

Thus, Fat Studies is not only about the body; it is about the Subject. Rosi Braidotti, whose feminist poststructuralist ethics "is not confined to the realm of rights, distributive justice, or the law, but...rather bears close links with the notion of political agency and the management of power and of power-relations," is useful for thinking through Fat Studies' insistence upon agency as part of its political project.[26] In fat activism (as distinct from Fat Studies, though they often overlap), a discourse privileging legal rights often prevails.[27] Fat Studies, perhaps without even realizing it does so, takes a page from the feminist poststructuralist playbook, where "responsibility is understood through issues of alterity, otherness, and difference as opposed to intentionality, action, behaviour, or the logic of rights."[28] Fat Studies focuses on the ways in which fat bodies are governed, whether that governing is done formally by the law or perhaps more subtly through language, representations, and rhetoric.

Fat people have much in common with others whose bodies are received as transgressive, problematic, or dangerous—people with disabilities, queers, old folks, racial and ethnic minorities, and so on—and whose subjectivity and social lives are thus compromised. These commonalities have led to many fruitful linkages between fat studies and other fields that investigate the relationship between subjectivity and various forms of embodiment: disability studies,[29] women's and gender studies,[30] and even food studies.[31]

FAT STUDIES AND THE PROBLEM OF INTERDISCIPLINARITY

In Joe Moran's intriguing book *Interdisciplinarity*, the index contains only three page ranges for "the body"—one for mind/body split, one for feminism and the body, and one for queer theory and the body. This speaks volumes about where one imagines the locus of scholarly interest in the body to be: either you are a centuries-dead White philosopher elevating the mind over the body, or you are a contemporary queer and/or feminist theorist resurrecting attention to the supposedly weak side of this equation while abnegating its force as a determinant of gender or sexuality. To wit: "The recent interest in the body in the humanities and social sciences, and particularly in feminist theory, is partly an attempt to subvert the traditional disciplinary division that has conceded this area of study to the sciences, particularly medicine and biology, leaving the supposedly autonomous products of the mind to the non-sciences."[32]

Fat Studies, like Women's Studies, Queer Studies and a host of other fields (Disability Studies, Critical Race Studies, etc.), is obviously interested in the body and embodiment—not only to do the work of disciplinary subversion outlined by Moran, but also because of their shared view of the body as a site of contested knowledge. Yet, while there are connections, there are also points of contention about embodiment among interdisciplines.

THE BODY IN FAT ACTIVISM AND HEALTH AT EVERY SIZE

Annette Kolodny locates the roots of women's studies in activism—acts of resisting subordinate roles in social movements—that morphed into consciousness raising that enabled analysis and political understanding.[33] Like Women's Studies and its counterpart in feminist activism, Fat Studies has its on-the-ground complements in fat activists and Health at Every Size (HAES) proponents. These entities are surely not mirrors of one another (the "theory" side and the "practice" side)—in fact, their interfaces are marked by predictable tensions and slippages. Elena Levy-Navarro, a Fat Studies scholar, recognizes the activist component of scholarly work in our area and extols the virtues of identifying with fatness not to make fatness a static ideal, but to revolt against those who value only life. "Instead, we would align ourselves with the queer against the straight, with death against a narrowly conceived life, and with a history that offers a pattern for a present that is only now partially conceived."[34] Just as Fat Studies scholars do, their counterparts in fat activism and HAES attempt to reposition fatness in fields of meaning and power.

A recent discussion on the Fat Studies listserv illuminated the overlaps and tensions among Fat Studies, fat activism, and HAES advocacy in their imaginings of embodiment. One list member articulated her concern that Fat Studies was too easy to write off because she felt that scholars in the field failed to engage adequately with facts about the body. She wrote,

> I value Fat Studies perspectives for offering skepticism of long-held beliefs about bodies and health, because the beliefs have driven the inquiry. However,

if **we** ignore or dismiss data that contradicts what we argue about bodies and health, rather than engaging with the methods used to obtain that data and the motives that may affect how that data was evaluated, etc., we will not gain credibility. In other words, if it is a foregone conclusion that anyone doing fat studies will automatically reject data that contradicts the idea that it is possible to be healthy at any size, etc., then we will not gain a foothold as an interdisciplinary field in academia.[35]

This poster went on to voice concern that Fat Studies had become too conflated with fat pride and HAES as political movements, "doomed to be criticized as the indoctrination of students into a new form of 'identity politics,' rather than valued as a real interdisciplinary point of critical inquiry."[36] The post reveals a fair amount of anxiety over work done in a field like Fat Studies where discourse about the body in an atmosphere marked by subjugation is necessarily politicized.

Much interesting discussion ensued from this post. List moderator Marilyn Wann argued that each of these three communities presents its own challenge to weight-based prejudice and discrimination, rejecting the poster's claim that the politicization of fatness undermines the work of Fat Studies because of some misguided belief that to be political is to be ignorant of data.[37] Many fat activists and HAES proponents jumped in to say that their perspectives and advocacy were based precisely on a skeptical interpretation of data about the fat body—that it was well thought-out, based on data and experience, and that led them to want to politicize fat embodiment, rather than some low-level knee-jerk reaction against science and medicine. Whatever their differences, the activist and advocacy versions of Fat Studies share a common complaint about how facts about "the fat body" are presented and to what social and political uses those facts are put.

CONCLUSION: FUTURE DIRECTIONS

At the outset of this chapter, I mentioned my colleague's cynicism regarding what he saw as a link between identity—not mine, per se, but *any* identity—and areas of study. A bulwark of one of those "timeless," "universal," "important" disciplines, it seems that he resists area studies in general, suspicious of their political genesis and political work. And he is surely not alone. Yet a generation (and more) of innovative scholarship in Women's Studies, Queer Studies, Critical Race Studies, Disability Studies, and other areas has set the stage for the emergence of Fat Studies as a viable field of inquiry that engages questions of embodiment while addressing larger social, economic, and political phenomena.

Personally and professionally, I have taken up Fat Studies as a way of better understanding human subjectivity. What started as an intellectual grad school fling with philosophers and social theorists like Hans-Georg Gadamer, Jurgen Habermas, and Michel Foucault, whose concerns with subjectivity were provocative yet abstract, was concretized for me a decade later when I started thinking about the questions they raised in conjunction with the lived experiences of fatness. But understanding human subjectivity is not the only game in town; other scholars are using Fat Studies to both ask and answer a number of vital questions about law and the self-governing citizen, the viability of social movements in a "postidentity" era, the meaning of embodiment in a technological age ever more marked by the experience of disembodiment, the meaning of risk in the context of biomedicalization, and the ability of science to produce and define social meanings. And so the list goes on.

As research in Fat Studies in the last ten years has taken up the problem of subjectivity/agency with increasing urgency, future work in the field should find us engaging more critically with the cultural and political meanings attached to such agency—complicating our comprehension of the relationship among body, embodiment, subjectivity, agency, and cultural meaning. We might also explore further the question of whether fat bodies themselves are agentic and disruptive, apart from their status as partially constitutive of fat *people*. Such considerations stand to contribute much to intellectual discourses on embodiment and larger sociopolitical contexts.

NOTES

1. Susie Orbach, *Fat Is a Feminist Issue* (New York: Berkley, 1978/1990), back cover.
2. Marcia Millman, *Such a Pretty Face: Being Fat in America* (New York: W.W. Norton, 1980).
3. Lisa Schoenfelder and Barb Wieser, eds., *Shadow on a Tightrope: Writings by Women on Fat Oppression* (San Francisco: Aunt Lute, 1983/1995).
4. Subjectivity, as I use the term throughout this chapter, is not quite the same thing as identity. Subjectivity entails agency and is a grounding for citizenship. Following Donald E. Hall, "one's identity can be thought of as that particular set of traits, beliefs, and allegiances that, in short- or long-term ways, gives one a consistent personality and mode of social being, while subjectivity implies always a degree of thought and self-consciousness about identity." Donald E. Hall, *Subjectivity: The New Critical Idiom* (New York: Routledge, 2004), 3.
5. Gail Weiss and Honi Fern Haber, "Introduction," in *Perspectives on Embodiment: The Intersections of Nature and Culture*, eds. Gail Weiss and Honi Fern Haber (New York: Routledge, 1999), xiv.
6. Ibid., xiii.

7. Gail Weiss, "The Abject Borders of the Body Image," in *Perspectives on Embodiment*, eds. Weiss and Haber, 52.

8. Stephanie Springgay and Debra Freedman, "Introduction," in *Curriculum and the Cultural Body*, ed. Stephanie Springgay and Debra Freedman (New York: Peter Lang, 2007), xxiv.

9. Ibid., xxvi.

10. Joe Moran, *Interdisciplinarity* (New York: Routledge, 2002), 107.

11. Marilyn Wann to Fat Studies Yahoo Group, "Fat Studies and the Body," December 31, 2008, http://groups.yahoo.com/group/fatstudies/.

12. Elena Levy-Navarro, *The Culture of Obesity in Early and Late Modernity: Body Image in Shakespeare, Jonson, Middleton, and Skelton* (New York: Palgrave Macmillan, 2008), 1.

13. Glenn A. Gaesser, *Big Fat Lies: The Truth about Your Weight and Your Health* (New York: Fawcett Columbine, 1996), 13.

14. Gina Kolata, *Rethinking Thin: The New Science of Weight Loss—and the Myths and Realities of Dieting* (New York: Farrar, Straus and Giroux, 2007), 223.

15. Eric Oliver, *Fat Politics: The Real Story behind America's Obesity Epidemic* (New York: Oxford University Press, 2006); Abigail C. Saguy and Kevin W. Riley, "Weighing Both Sides: Morality, Mortality, and Framing Contests over Obesity," *Journal of Health Politics, Policy and Law* 30, no. 5 (October 2005): 869–921; Paul Campos, *The Obesity Myth* (New York: Gotham, 2004); Michael Gard and Jan Wright, *The Obesity Epidemic: Science, Morality and Ideology* (New York: Routledge, 2005).

16. Wann, December 31, 2008.

17. Cheryl L. Nicholas, "Disciplinary-Interdisciplinary GLBTQ (Identity) Studies and Hecht's Layering Perspective," *Communication Quarterly* 54, no. 3 (August 2006): 308. See also Michael Hecht, "2002—A Research Odyssey: Toward the Development of a Communication Theory of Identity," *Communication Monographs* 60 (1993), 76–82.

18. Ibid., 318.

19. Angie Morrill to Fat Studies Yahoo Group, December 29, 2008, http://groups.yahoo.com/group/fatstudies/.

20. Thomas J. Csordas, "Embodiment and Cultural Phenomenology," in *Perspectives on Embodiment*, eds. Weiss and Haber, 147.

21. Ibid., 149.

22. Ibid.

23. Carolyn DiPalma, "Body Politics: Webs of Embodiment, Medicine, Science, Technology, Nature and Culture," *Theory and Event* 6, no. 2 (January 2002): not paginated—accessed via ILL online printout.

24. John T. Sanders, "Affordances: An Ecological Approach to First Philosophy," in *Perspectives on Embodiment*, ed. Weiss and Haber, 121.

25. Lauren Berlant, "Slow Death (Sovereignty, Obesity, Lateral Agency)," *Critical Inquiry* 33 (Summer 2007): 754–780; Elspeth Probyn, "Silences behind the Mantra: Critiquing Feminist Fat," *Feminism and Psychology* 18, no. 3 (2008): 401–404; Antronette K. Yancey, Joanne Leslie, and Emily K.

Abel, "Obesity at the Crossroads: Feminist and Public Health Perspectives," *Signs: Journal of Women in Culture and Society* 31, no. 2 (2006): 425–443.

26. Rosi Braidotti, *Transpositions* (Cambridge, UK: Polity Press, 2006), 12.

27. Anna Kirkland, "Think of the Hippopotamus: Rights Consciousness in the Fat Acceptance Movement," *Law & Society Review* 42, no. 2 (2008): 400.

28. Stephanie Springgay, *Body Knowledge and Curriculum: Pedagogies of Touch in Youth and Visual Culture* (New York: Peter Lang, 2008), 34.

29. Charlotte Cooper, "Can a Fat Woman Call Herself Disabled?" *Disability & Society* 12, no. 1 (February 1997): 31–41; April Herndon, "Disparate But Disabled: Fat Embodiment and Disability Studies," *NWSA Journal* 14, no. 3 (Fall 2002): 120–137; Anna Kirkland, "What's at Stake in Fatness as Disability?" *Disability Studies Quarterly* 26 (2006); available online at http://www.dsq-sds.org/2006_winter_toc.html.

30. Susan Koppleman, ed., *The Strange History of Suzanne LaFleshe: And Other Stories of Women and Fatness* (New York: Feminist Press at CUNY, 2003); Jana Evans Braziel and Kathleen LeBesco, ed., *Bodies out of Bounds: Fatness and Transgression* (Berkeley: University of California Press, 2001); Orbach, *Fat Is a Feminist Issue.*

31. Julie Guthman, "Can't Stomach It: How Michael Pollan et al. Made Me Want to Eat Cheetos," *Gastronomica* 7, no. 2 (Summer 2007): 75–79; Alice Julier, "The Political Economy of Obesity: The Fat Pay All," in *Food and Culture: A Reader*, 2nd ed. eds. Carole M. Counihan and Penny Van Esterik (New York: Routledge, 2007), 482–499.

32. Moran, *Interdisciplinarity*, 106.

33. Annette Kolodny, " 'A Sense of Discovery, Mixed with a Sense of Justice': Creating the First Women's Studies Program in Canada," *NWSA Journal* 12, no. 1 (2000): 144.

34. Levy-Navarro, *Culture of Obesity*, 33.

35. Sonya Brown, personal communication re. "Fat Politics across the Spectrum" on fatstudies@yahoogroups.com, January 10, 2009.

36. Ibid.

37. Marilyn Wann to Fat Studies Yahoo Group, "Was Fat Politics/Response to Sonya," January 21, 2009, http://groups.yahoo.com/group/fatstudies/.

CHAPTER 7

IDENTITIES WITHOUT BODIES: THE *NEW* SEXUALITY STUDIES

LISA JEAN MOORE AND LARA RODRIGUEZ

PROLOGUE: DICK IN A VICE

During graduate school in the early 1990s, one of us (Lisa) worked for a community-based organization called San Francisco Sex Information (SFSI). After fifty hours of training, volunteers worked shifts on a phone line answering callers' queries about sex. The training was truly interdisciplinary as it covered many topics—for example, sexual health, sexual identity, and genital anatomies—as well as phone etiquette, including extensive instruction in nonjudgmental responses. Volunteers were taught that the most common questions asked would be about average penis size, location of the clitoris, normal frequency of masturbation, and places at which to be tested for STDs. Volunteers also learned how to manage misinformed callers who thought that SFSI was a phone sex service and to detect callers who were attempting to engage in nonconsensual phone sex from the volunteer. For example, trying to get a volunteer to say the word "panty" was one technique devious callers used.

After Lisa's training, she selected a shift on Monday evenings and manned the station with two other volunteers and a shift supervisor. Armed with her nonjudgmental answers, such as "between 5–7 inches" and "as long as it doesn't interfere with your activities of daily living, then it can be considered normal," she was ready for her first call:

> *LJM: "San Francisco Sex Information."*
> *Caller: "Hi, I have a question. My professional dominant does ball in vice torture and I was wondering if it will affect my sperm. She does it for about 20 minutes at a time and I don't want to mess up my sperm."*

Lisa could not have been more unprepared. But quickly remembering the training, she repeated the question aloud for the shift supervisor to hear. He nodded, pulled from the bookcase a volume of photographs of sexual practices and "torture devices" and opened up to a color glossy of an erect penis with a strap secured around the testicles and a rod protruding from an apparatus designed to tighten the straps. The visual did not give Lisa any more insight into answering the caller's question and, if anything, just further distracted her from the nonjudgmental sex-positive place in which she was supposed to be. She answered by asking the professional dominant to modify the practice to help protect sperm production, which may or may not have been accurate.

This reflection, deeply reminiscent of late twentieth-century San Francisco's lesbian, gay, bisexual, transgender, queer (LGBTQ) communities and AIDS activism, today seems almost a quaint historical illustration of sexuality studies fieldwork. Those were the good old days when investigating the "deviant practices" of "subcultures" was at the cutting edge of the field. In discourses of HIV/AIDS, at that time an acute and lethal disease in the United States, certain bodily practices trumped sexual identity as a means to make sexual risks intelligible. At SFSI, we asked if someone was an anal-insertive partner or an anal-receptive partner as part of our standard stream of risk assessment questions. This seems somewhat "retro" now after so many changes in how we access information and interpret sexuality. And while sexuality studies today still combines the scholarship of psychology, anthropology, sociology, history, public health, medicine, law, politics, literature, art, and philosophy, something rather dire seems to have happened to the *fleshiness* of sexuality studies. So we ask: If Oprah and Tyra can dedicate entire episodes of their shows to "vajayjays" (vaginas), and images of Britney Spears's peek-a-boo labia fly across the Internet, why is sexuality studies now so fleshless and disembodied? Where have all the bodies gone?

SEXUAL IDENTITIES DESPITE BODIES

When considering twenty-first-century sexuality studies, embodiment and bodies appear to have receded behind identity and power as core variables of interpretation and analysis. The man who had his "dick in a vice" playing with a professional dominant has given way to discussions of his identity as a "john" or "sexual deviant" and of the regulatory apparatus that "produced" his identity as a "person of interest" vulnerable to surveillance. Although some scholarly work details and charts the variety of sexual behaviors included in S/M practices, more often articles focus on the development of identities and subcultures. Take, for example, "The Time of the Sadomasochist: Hunting with (in) the 'tribus'" or "A Hermeneutic Phenomenological Investigation of the Construction of Sadomascochistic Identities." Academia seems to favor Foucault over fist-fucking, but amid all this intellectual masturbation, we are hardly blushing. Where are the verdant, fleshy, leaky, messy, drippy, carnal descriptions of bodies as front and center in sexuality studies? Why have they disappeared in favor of the sign and signifier language that distracts from the more phenomenological or robust descriptions of the body?

If we flash forward from the 1990s to 2009, coauthor Lara's undergraduate experiences also provide some insight into sexuality studies, on the ground. Her boyfriend's little sister recently asked Lara, "Do you consider yourself bisexual?" Lara is somewhat puzzled by the sixteen-year-old's question, but even more so by the girl's willingness to claim for herself a bisexual identity while admitting, "guys have it so easy, their body is so easy to figure out" (in reference to masturbation). How had this teenage girl arrived at a sexual identity *before* arriving at an intimate understanding of her own body and its workings? And why should the male sexed body en masse appear to her so much "easier to figure out" than her own? We investigate here how the new sexuality studies (1990s–present) in part reflects as well as contributes to a cultural consumer practice that demands identifications in/by predetermined groups before we even know our anatomies, pleasures, turn-ons, and turn-offs. But how has this new sexuality studies been generated?

It might be important to site (and cite) where sexuality studies exists institutionally and academically—it resides somewhere between women's or gender studies programs and lesbian, gay, bisexual, transgender (LGBT) studies. As faculty work to teach students, there is often a direct relationship between what happens in the classroom and what happens in the dorm room. Across several American undergraduate institutions, most of Lara's cohort (i.e., gay/straight/bi/trans men and women and students concerned with sexuality and gender) discusses sex, but their understanding of each other's sex lives is framed always by identity. In other words, these conversations seldom reveal the intimate details of *how* one's peers have sex, but rather focus on *who* the sex is with. Frequently these conversations concerning sexual histories and desires go something like this: "She says she's a lesbian, but she's FUCKING a dude!" or "I might be gay, but I would love to eat pussy," or, "But you're a lesbian! Imagining a dick in you just seems wrong." As Lara was writing this chapter, a twenty-two-year-old woman who identifies as a butch lesbian in Brooklyn reveals that when she goes to the gay bar, she is frequently hit on by gay men! "They all think I'm a transman now, not a lesbian."

So who's queer now? The anxiety, celebration, inclusion, and proliferation of inconsistent identities, and subsequently sexual practices, suggest that sexuality studies could be blossoming. As exciting as the rise of queer and trans boundary imploding identities might be, we as female-bodied feminist scholars feel somewhat estranged and nervous. Yet perhaps it was not quite so sudden. Maybe the accomplishments and changes made through sexuality studies via the efforts of deconstructionism have been until now unidirectional. We, as others, are writing from a politics of deconstruction, too, so we may shift the field's current preoccupation with identity toward specific embodied identities that have thus far been more difficult to consider.

In 2005, Linda Garber's "Where in the World are the Lesbians?" anticipated the cultural shift. She asked, "How come every time 'queer studies' looks in a new direction it seems to reproduce the same 'male homosexual studies'?"[1] We echo this question as we observe in academic journals a marked neglect of women's sexuality and the embodied issues regarding female-bodied persons. The twenty-first-century sexuality studies canon does not seem to be interested in reproductive politics, abortion, or even the vital politics surrounding the HPV vaccine.[2] However, we are inundated with articles concerning queer migration and global masculinities: "Dude-Sex: White Masculinities and 'Authentic' Heterosexuality among Dudes Who Have Sex with Dudes" (from *Sexualities*), or "Constructing the Neoliberal Sexual Actor: Responsibility and Care of the Self in the Discourse of Barebackers" (from *Culture, Health and Sexuality*). The abundance of interesting identities with respect to male sexuality—its exclusivity, its ability to eclipse other identities, and its lack of engagement with the body—concerns us. If scholars prioritize and privilege identity formation over embodied experience, who will confront the messy, complicated, consequences of living within bodies that are inevitably corporeally gendered?

In the remainder of this chapter, we address this tension present within the new sexuality studies as suggested by the erasure of a fleshed, sexed body. To make any claims about the new sexuality studies qua academic discipline, we performed a content analysis of seven peer-reviewed academic journals and two self-named "sexuality studies" readers. We use this data to illustrate the ways in which the body is present or absent in the field. After examining four areas in which (and how) the body is defined, we then examine burgeoning areas where the body becomes (is becoming more) visible. Finally, we provide a "reading" of what is not seen—which bodies and their parts have not yet been taken up as worthy of study in the new sexuality studies.

From our vantage points in the academic worlds of professional scholarship and undergraduate coeducational living, it appears that individuals are tempted by claiming the security of an identity (or set of identities). And yet these very identities can limit our imaginations and reduce complexity. We do not mean to suggest that identities are useless. The ability to identify is powerful as it enables humans to concisely express not only who we are (and are not), but also to focus and reveal our affinities, histories, politics, cultures, and values. We can be part of a tribe and different from other tribes; we can know our social roles and adopt social scripts; we can follow socially regulated norms guided by our peers and the media we consume. In this context, the Kinsey scale and its popularization in the 1950s and 1960s presented us with a notion of sexual identity as far more complicated than our language and terminology would allow. And yet, we desperately cling to the notion of singular or uniform identities. We suggest this is because "who we are" depends very much on whom we say we are fucking. If we wish to marry our practices to our selves, identity becomes the symbolic marker for both who we are and what (and who) we do.

Sexuality studies has been successful insofar as it has demonstrated that discourse has value, but we argue that the discursive production and proliferation of sexual identities and subcultures may be too dismissive of flesh and bone, oozy and carnal bodies. To wit, how can we as so-called bi, gay, lesbian, straight, feminist, trans men and women understand our own sexualities outside of identity-laden terms and movements that have fought so hard to give us self-understanding, voice, and stature? As social scientists and sentient humans, it is obvious, at least to us, that there is discrimination and a profound imbalance of power. Sexual power and agency are often distributed through matrices of heteronormativity, male domination, and white privilege within the capitalist political economy. For example, with respect to Proposition 8 (the California voter initiated proposition to overturn the Supreme Court ruling on same sex marriage), sexual identity matters because it enables us to speak from a political position. However, as embodied sexual subjects, our bodies also matter because it is our corporeality, and not exclusively our identities, through which we can derive some pleasure, pain, exquisite suffering, and occasional transcendence—sometimes despite our identities. How then might sex be conceptualized within what we are calling the *new* sexuality studies such that our bodies are as present and vital as our sexual politics, histories, oppressions, and identities?

We paraphrase Lisa Jean Moore and Mary Kosut's definition of the body to suggest that the sexual body is a fleshy, verdant, carnal, sensate, drippy, leaky, and engaged organism that is composed of bones, blood, organs, hormones, and fluids, as well as statuses, hopes, fears, and anxieties.[3] It is this sexual body that is implicated, either entirely or partially, within the experiences of sexual interaction. Since we both identify as feminists, we are curious about the ways in which the fleshiness of bodies, particularly vulnerable bodies, female bodies, and children's bodies, may be marginalized in the epistemological record and project of sexuality studies.

Simultaneously, in certain areas of sexuality research—those areas more concerned with surveillance and intervention—particular bodies are seen as troublesome. There is a distinction between how the body gets displaced by identity politics (e.g., gay rights) and how the body was relegated to a back seat by Foucauldian/historical narratives about the social. The first displacement is an historical artifact; the second was a theoretical mistake. We argue for a commitment from new sexuality studies to continue theoretically sophisticated scholarship, while allowing the body, not exclusively female bodies, to remain front and center.

EVERYTHING OLD IS NEW AGAIN

One of the most significant complexities about the field of sexuality studies is that it is widely interdisciplinary, including sexologists, sociologists, queer theorists, public health scholars, historians, sex therapists, health educators, endocrinologists, anthropologists, feminist theorists, activists, lawyers, and philosophers. So what is the new sexuality studies? As described in a text that attempts to define the canon, *Introducing the New Sexuality Studies*, "The new sexuality studies understand sexual identities as historically emergent...the new sexuality studies perspective does not deny the biological aspects of sexuality. There would be no sexuality studies without bodies...it is this deep view of sex as social that we hope to convey in this volume."[4] Sexual identities are historically emergent; sexuality studies must be embodied, and it must also understand sex as social—in our content analysis we found that this definition is more a wish list rather than an actual scholarly practice. In attempting to periodize sexuality studies so as to launch a critique, we here characterize the temporal shifts in sexuality studies as a means to enable us to see what is "new" or at least as "revisioned" over time. Clearly, as with most historical shifts, these are not neat or bounded time periods, rather there is overlap and bleeding between and among different eras.

We begin with the 1950s as the start of the previous generation of sexuality studies. This era relied more heavily on quantification and measurement with an emphasis on Kinsey-esque sexology surveys to measure sexual behavior and a reliance on statistical tools of "objectivity" to present nonbiased statistical aggregates within the scientific method. Masters and Johnson also used clinical observations to document actual behaviors and psychodynamics of sexual behavior and stimulus. Hormones, sizes, shapes, and actions were all quantified and given specific meanings and interpreted for larger sociocultural meaning. This era, when seen through the more updated lens of queer theory and critical sexuality studies, can be read as a prescriptive science that constructs the normative (and normates) and provides scientific foundations for why certain (i.e., heterosexual) behaviors are prevalent.

In the 1970s, sociologists John Gagnon and William Simon offered an analysis of sexual scripts beginning with the discourse of social constructionism. According to sociologists Michael S. Kimmel and Rebecca F. Plante, they provided "the most important theoretical breakthrough in thinking about sexuality."[5] Following Gagnon and Simon, there was a shift away from essentialist and biomedical discourses of the body. Sexuality studies seemed to come into its own apart from sexology, that is the scientific study of sex. In regard to sexology, "the aim of this new science was to study the sexual life of the individual within the context of medicine and the social sciences."[6] Janice Irvine suggests that "the tensions between social and biological sciences in sexology have always been a problem—gradually sexology leaned more and more toward biology and medicine in an attempt to legitimate the field as a science."[7] Within sexuality studies, sexology appears to be recognized as a discourse of pathology and oppression via the medical gaze while the *new* sexuality studies is a discourse of deconstruction and social change. Anthropologist Gayle Rubin writes, "Concepts of sexual oppression have been lodged within that more biological understanding of sexuality."[8] In her view, the body appears as part of *that more biological understanding* that holds us back from engaging a radical theory of sex.

The study of human sexuality in the 1970s changed markedly as a result of important social movements such as the Gay Liberation and Women's Movements. Specifically, the Women's Self-Help Health Movement created venues for new explorations of sexual topics previously invisible. In the wake of social change and controversy, issues such as female ejaculation, clitoral form and function, menstrual extraction, abortion, birth control, and menopause were finding their way into sexuality studies as the discipline emerged. Influenced by the rise of second-wave feminism, the 1980s brought to the fore contested issues of sex work, pornography, and sadomasochism. However, it seems that sex and bodies belonged to those in the marginalized identities of sex worker, LBT identities, and Bondage, Domination, Sadism and Masochism (BDSM) practitioners. Within sexuality studies heterosexuality, adolescent girl bodies, or so-called normative practices were not given equal attention. In 1984, Gayle Rubin announced, "A radical theory of sex must identify, describe, explain, and denounce erotic injustice and sexual oppression...it must build descriptions of sexuality as it exists in society and history."[9] Rubin cited sexual essentialism as having dominated more than its fair share of sexualities discourse and although she claims sexology and sex research may prove valuable as they "can provide an empirical grounding for a radical theory of sexuality," her investment and work exists consistently within a social-political framework.[10]

The HIV/AIDS epidemic framed sexuality studies throughout the late 1980s and into the mid-1990s, popularizing the field of queer theory. We will use the time since HIV/AIDS to signify the new sexuality studies. That is not to suggest that HIV/AIDS is not still relevant to sexuality scholarship; however, the emergence of AIDS and the innovation of identity as a category highly relevant to "old" sexuality studies appear to have been significant in guiding the research. We attribute the 1990s as beginning to shift to a newer sexuality studies in part because of the production of AIDS as an "epidemiological disaster." Disastrous in its viral replication and physiological effects, it destabilized any notion of a sociological or empathetic empiricism by existing as a phenomena both invisible and hypervisible. Under the weight of the AIDS crisis, the unreliable, easily infiltrated fragile flesh we call our body, susceptible to mutation, turns queerly and problematically within scholarship as the body without organs. Following from Mary Douglas, we too see bodies as symbolic of purity and danger. In Western culture, human bodies are dirty, disgusting, and associated with guilt. These meanings have trickled into the field of sexuality studies, often discounting it as an academic discipline. In other words, notions of the dirty, diseased, epistemologically disastrous AIDS body may have scared people off from talking about visceral embodiment, and as a result scholars have looked for more abstract and highbrow ways to talk about sex and sexuality.

Notably, the two textbooks from the new sexuality studies we surveyed were sparse in their attention to HIV/AIDS. It is ironic, in our analysis, how the integration of AIDS into sexuality studies seems to have moved the field farther away from the body/flesh and more into identity.

METHODOLOGY

We honed our methodology using ideas from the recent book *Missing Bodies: The Politics of Visibility*, which served as a guide for revealing/exposing the missing bodies from the field of sexuality studies. As sociologists Casper and Moore write, "Using our emergent ocular ethic—so necessary and vital in this age of surveillance and security—we focused our attention on excavating hidden, concealed, and buried bodies, and on revealing the found objects of our investigatory practices. But in order to consider these found objects—such as children's agency or a baby's health—we needed to dissect and display the structural and symbolic processes that led to their social erasure."[11] In addition to coding our data for what types of bodies are visible in the scholarly field of sexuality studies, we must also consider what bodies get eclipsed. The question of *what* is deemed worthy of study, publishable, interesting, scholarly, significant, important, is also often the question of *who* is considered worthy of study, publishable, interesting, scholarly, significant, and important. Whose bodies (and practices) matter?

To map the terrain of the new sexuality studies, we collected and analyzed data from seven interdisciplinary, English-language, peer-reviewed journals and two contemporary edited collections of key readings in sexuality studies. Although this is a limited group of journals and scholarly books, and thus a partial interpretation of the field, we have attempted to triangulate our analysis through content analysis as well as grounded theory of key codes that emerged from our coding operations.[12] These journals are accessible through the State University of New York library system. The texts were books that have been commonly assigned based on a web-based sample of twenty syllabi for sexuality and gender studies courses over the past three years. Articles were analyzed within the dates available as indicated on the table.

Through our analysis, we established a sentence to sum up the types of scholarship of the journal and thus created an identity for each journal with respect to its treatment of the human body.

1. *Annual Review of Sex Research*—Sexuality discussed in terms of psychological and psychosomatic realities. As bodies are unavoidably corporeal, they are subject to pathology, disease, abuse, rape, dysfunction, disorder, and subsequently, medical intervention.

2. *The Journal of Sex Research*—Examines sexuality as behavior and identity formation within particular local and specific social/cultural contexts across different demographics within North America.

3. *Culture, Health & Sexuality*—Sexuality as always performed and located within specific local situated cultural contexts—depending on the context/community, some bodies are more "AT RISK" (referring to drug users and unprotected sex) than others and therefore vulnerable to STDs/STIs, and threats to the population.

4. *Sexuality & Culture*—Examines intersections between sexuality and culture—particularly it is interested in media representations and the ways in which culture is integrated into our sexual imagination; sexuality is a social negotiation and its practice coincides with how information is filtered/trafficked through media/popular images.

5. *Sexualities*—Sexual identities are personal and political—sexuality is a playful experience and need not be just a discourse of oppressions. Sexuality is not just a practice and behavior; sexual identities give rise to unique, diverse sexual subcultures, which contribute to the cultural imagination.

6. *GLQ*—This journal explores the history of sexualities and sexual subcultures; sexuality breeds and contributes to high culture, history, art, music, and literature; sexual identities are a negotiation and product of historical, artistic, legal, and political representations.

7. *Journal of History of Sexuality*—Sex exists within culture, traditions, beliefs, practices, geographies, governments, politics, and temporalities; suspicious and critical of biomedical/essentialist discussions of sexuality.

Name of Journal	Inaugural Year * Available for our analysis	Disciplines	Demographic	Count of sexual subject
Annual Review of Sex Research	1990–present [*1990–2006]	Biomedical Psychology Sexual Health Sociology	Women, Men, Children	(dys)function, disorder: n=11 response, arousal, orgasm: n=8 rape/abuse: n= 8 AIDS/HIV: n=6 children: n= 6 reproduction: n = 5 PMS, menopause: n = 5 medicalized language: n = 25 gender: n = 8 lesbian: n = 2 gay: n = 3 hetero: n = 4 identity: n=3 behavior: n = 23 AIDS/HIV n = 6 feminism: n = 3
The Journal of Sex Research	1965–present[* 1975–present]	Sociology Anthropology Biomedical/ Psychology	Crosses age, class, occupation, sexuality, gender, race	adolescent/young: n = 22 heterosexual: n = 17 homosexual: n = 29 gay: n = 28 lesbian: n = 17 reproduction: n = 1 pregnancy n = 2 identity n = 8 gender: n = 33 trans: n = 9 risk: n = 14 AIDS/HIV: n = 13 birth control: n = 3 condom: n = 11 behavior: n = 44 feminist: n = 7 queer: n = 2 scripts, n = 8 body: n = 6
Culture, Health, & Sexuality	[1999–present]	Anthropology Sociology Sexual Health and Behavior	homosexuals, MSM, heterosexuals, drug users,	HIV/AIDS: n > 80 body: n = 2 gay/male homosexuality: n = 40 MSM: n = 10 lesbian: n = 7

Continued

Name of Journal	Inaugural Year * Available for our analysis	Disciplines	Demographic	Count of sexual subject
				gender: n = 20 identity: n = 9 contraception: n = 3 condom: n = 18 trans: n = 4 queer: n = 2 reproduction: n = 8 pregnancy: n = 6 fertility: n = 2 scripts: n = 4 practice: n = 8 rape: n = 3 homophobia: n = 5 violence: n = 5
Sexuality & Culture	1996–present [*2000–present]	Sociology Cultural Studies	Consumers of culture, West	homosexual: n = 0 gay: n = 13 heterosexual: n = 5 lesbian: n = 5 identity: n = 6 gender: n = 13 scripts, n = 4 porn: n = 14 pedophilia/child abuse: n = 10 victimization/rape/ violence: n = 13 behavior: n = 5 trans: n = 2 queer: n = 5 body: n = 2
Sexualities	[*1998–present]	Gender Studies GayLesbian Studies Cultural Studies (subcultures)	bodies without organs, focus is sexual/ political identity, homosexual, queer, gay, lesbian, butch, feminist	queer: n = 16 homosexual/gay: n = 40 lesbian: n = 30 feminism: n = 10 identity: n = 23 hetero/straight: n = 22 homophobia: n = 8 trans: n = 9 practice: n = 9 gender: n = 16 body: n = 4
GLQ	[*1997–present]	Gay, Lesbian Studies Cultural Studies Humanities	GLBTQ	gay/homosexual: n > 60 lesbian: n = 35 feminism: n = 8 trans: n = 26

Continued

Name of Journal	Inaugural Year * Available for our analysis	Disciplines	Demographic	Count of sexual subject
				queer: n > 70
				bisexual: n = 4
				straight/hetero: n = 8
				gender identity/roles: n = 25
				civil rights, activism: n = 12
				antigay/homophobia: n = 9
				HIV/AIDS: n = 10
				Body/flesh: n = 10
Journal of History of Sexuality	[*1990– present]	Humanities History, Art, Cultural Studies	N/A	bodies: n = 31
				gender: n = 37
				literary: n = 45
				queer: n = 34
				prostitution: n = 25
				porn: n = 11
				homophobia: n = 6
				rape: n = 8
				STIs: n = 11
				HIV/AIDS n = 12
				trans: n = 7
				lesbian: n = 35
				homosex: n =43
				gay: n = 35
				heterosex: n = 6
				masturbation: n = 6
				repro: n = 17
				abortion: n = 7
				birth control: n=2
				condom: n = 2
				feminism/feminist: n = 21
				masculinity: n = 28
				maternal/mother: n = 11

The two edited collections in our analysis were chosen due to the recent publication dates, the range of previously published staples of the field, and the diversity of disciplinary approaches. The first book is *Introducing the New Sexuality Studies: Original Essays and Interviews* containing sixty-five original, commissioned articles, essays, and interviews. The second book is *Sexualities: Identities, Behaviors, and Society* with forty-seven articles formatted as a mixture of previously published and original submissions. Combined, these books include 112 articles and interviews representing scholarship framing the new sexuality studies. Detailed content analysis was performed with notes including direct quotations, general observations, and summaries.

NO SEX, WE'RE ACADEMICS

The most significant and startling finding of our research on the textbooks was the lack of embodied description within the contents of the book. There were no images, no drawings, no photographs, and very few graphic descriptions of sexual acts. Very unsexy, if you will. Clearly there are financial and publishing limitations to what can be represented in an edited volume—however, we believe there might be something else going on.

Is the new sexuality studies a canon (a "cannon") without (female) organs? Ironically, the effect of Rubin's "radical" theory of sex (where gender and sex are separate) as realized within real world schema/practice present us with a problem between the discourse-subject. Who is most invested and responsible in maintaining each discourse respectively becomes, naturally, the subject of the discourse; Beasley observes,

> Feminist approaches typically focus on gender and usually highlight women's positioning whereas sexuality studies have produced a greater volume of attention to men and their sexualities, specifically to gay men's sexual positioning. If the main subject of Gender Studies to this day remains women, then the central subject of Sexuality Studies has been gay men. This focus occurred partly because of a residual traditional privileging of men's perspectives and partly because lesbians have sometimes seen themselves, and have been seen, in relation to women's experiences. In other words, lesbians have often been located under the umbrella of Feminist perspectives rather than as primarily described in relation to (homo)sexuality.[13]

This coincides with Rubin's argument that a theory of sexuality is apart and distinct from a theory of gender oppression, so women's health and sexuality (particularly that of lesbians) are perceived as gender/feminist concerns more so than sexuality concerns.

VULNERABLE PARTS, CLINICAL ANATOMIES

Within the contents of journals and books, human bodies are defined most frequently within four categories: *vulnerable entities, commodities, clinical anomalies,* and *anatomical parts.* We examine what these categories mean more specifically below. More importantly, certain bodies are consistently missing from the contents of these scholarly journals.

Specifically, there are no essays about children's bodies as sexual actors, and in particular with experiences narrated by the children themselves. Perhaps related, adolescent girls' sexuality appears very infrequently with perhaps a token piece once every few issues. Sexual health with respect to particular reproductive concerns of contraception and management of fertility is also absent from sexuality journals. Perhaps the absence of these types of bodies is due to the antiessentialist stance, which is critical of the medical gaze and interventions upon the sexed body. This stance has resulted in a less than sufficient exploration of the realities and practicalities of sexuality.

Vulnerable entities refer to bodies at risk with respect to HIV/AIDS and STIs as well as sexual violence. This is by far the most common way to describe and understand the body—it is an object that is potentially victimized or infected. A discourse of risk within the new sexuality studies has grown from the HIV/AIDS epidemic and the academic community's response to AIDS.

Clinical anomalies refer to scholarship about physiology and sexual or anatomical "dysfunction." Clearly this is also an indication of the dominance of discourses of medicalization rife within the field of sexuality studies. While there are dissident voices critical of the discourse of medicalization, the encroachment of allopathic, biomedical definitions over sexual bodies is prevalent within all the journals we analyzed. Bodies, as corporeal entities that contain people and define them, are not centrally considered in much of sexuality studies. However, when they are taken seriously it is often in the service of explaining and defining both bodily and identity dysfunction. Understanding the role of medicalization and the dominance of medical knowledge production over the realm of sexuality studies is partially responsible for the discourse of dysfunctionality. Erectile dysfunction ranks most highly in frequency in the literature about clinical anomalies; form and function, particularly that of phallocentric sexual activity, motivates much of the scholarship and research within the realm of biomedical and pharmaceutical research.

Beyond erectile dysfunction, other bodies are defined within the medicalized realm as *clinically* interesting or deviant. These, in order of their frequency, are inappropriately sexual bodies, sexual predators, masturbating bodies, intersexed bodies, and infertile bodies. Bodies are referred to as parts, rather than as a fleshy whole. For example, the most commonly discussed body part within the sample is the male penis; among the most notable, "All Hail the Penis" (*Culture, Health & Sexuality*). Scholarship examines how cultural understandings emerge about certain body parts and their sexualization. Body parts are symbols of larger social structures or social values to certain symbolic body parts.

TRANS BODIES

Susan Stryker and Steven Whittle suggest that trans studies emerges, "[I]n the 1990s, a new scholarship, informed by community activism started from the premise that to be trans was not to have a mental or medical disorder. This fundamental shift was built upon within academia and enabled trans men and women to reclaim the reality of their bodies, to create with them what they would, and to leave the linguistic determination of those bodies open to exploration and intervention. Trans studies is a true linking of feminist and queer theory."[14] What is interesting about trans studies is that as it emerges and is folded into new sexuality studies, it seems to be almost exclusively presented as an identity discourse with difficulty putting body front and center—is this a concern of privacy or a sense that the body is just a construction and that trans people are dupes of the system in wanting to change their bodies?

Heather K. Love writes, "If for a long time queer studies moved in the direction of ever-greater flexibility of its defining terms, it seems to have reached an internal limit as it confronted the new field of trans studies. Sexuality's resistance to essentialism, which once seemed so promising, has begun to seem less interesting in relation to the stubborn materiality of gendered embodiment."[15] Although we identify trans as a gender identity and GLB as sexual identities, they have been historically confused and mistaken for one another and continue to be studied under the larger umbrella of sexual deviance. Ironically, trans studies and narratives, especially those of transsexuals, have the potential to lead sexuality studies back to the body, ironically opposing the antiessentialist platform on which queer theory is based. This movement is not without anxious commentary. For example, Dean Spade admits, "I am cautious about using information about my own transition or body for fear of participating in an exercise whereby I am more or less 'real' depending on how much medical recognition and intervention I have undergone."[16]

Trans studies calls to our attention the dual interplay between gender and sexuality, and the need to understand how these discourses inform and interact, but do not run parallel. Just because there are transgender subjects inhabiting the contemporary, prestigious queer and antiessentialist location, does not mean that the material concerns of occupying transsexual bodies disappears. GLB identities in the United States are not subject to the same medical and legal mediation that transsexual identities are. As a result, GLB groups' political interests differ from those belonging to the trans community. How to reinstate and make significant the material somatic feelings while simultaneously asserting antiessentialism?

There is within sexuality studies a hierarchy of identity, and subsequently a hierarchy of knowledge. For example, Patricia Elliot, summarizing Jay

Prosser writes, "Some aspects of transsexual experience are irreconcilable to queer aspects that include: 'the importance of the flesh to self';...perhaps above all...a particular experience of the body that can't simply transcend...the literal."[17] Queer Theory, in its attempts to erase identity, actually multiplies it—identities proliferate, which is not in itself a negative outcome until particular embodied subjects and their material concerns are not accounted for or not deemed worthy of academic investigation. As Elliot argues in the case of transsexuals, "Subjecting transsexuals to the gaze of queer theory as Butler and Halberstam have done actually distorts their experience to the point of rendering them invisible...this subjection implies that the only politically valuable aim is the transgender undoing of categories of sex, accompanied by the celebration of the incoherent, the noncongruent and the unintelligible body."[18]

BRINGING THE BODY, EVERY BODY, INTO VIEW

Our analysis reveals that sexuality studies as a field has privileged a particular view of the body that either dissects (into parts), renders invisible (in terms of images), or reduces to categorizations based on identity. This way of seeing (or not seeing) the body negates the fleshy embodied experience of sex itself. We argue that the new sexuality studies must explicitly engage with bodies, and scholarship should be reflexive to reveal the complicated natures of sexual experience. Also, scholars should attend to how bodies are framed within larger cultural debates/media and within academia. Jeff Hearn suggests one way that the body could more fully occupy contemporary sexuality studies is via the development of "crip theory" which lies "at the intersection of disability theory and queer theory."[19]

Although conscious of former feminists' tendencies to "lump" sexualities discourses with those of gender, Stein is also wary of current academic as well as pedagogic trends that advocate for a separate treatment of sexuality from gender. She suggests academics engage in "self-reflexive lumping" of sexuality and gender so that we may understand these discourses as distinct, but also as sharing complicated intersections; in order to avoid a further fragmented identity politics, we should "try to gender sexuality and sexualize gender."[20]

Perhaps we should look to other interdisciplinary fields to continue to push sexualities studies in bodily directions: Science and Technology Studies, Disability Studies (particularly the work of Crip Theory), and Transgender Studies all have worked to explore what a techno-sexed body is in the twenty-first century. It may also be useful for sexuality studies to be less resistant to different theoretical modes of investigation; for example, finding our way back to the body under an operational essentialism or even skeptical empiricism may mean embracing a more phenomenological and affective course. Particular bodies that reside at the crossroads of medical/technological/reproductive/genetic intervention are increasingly coming into focus with limited commentary from sexuality studies. We call for scholarship that understands these bodies as raced, gendered, sexed, and sexual.

NOTES

1. Linda Garber, "Where in the World Are the Lesbians?," *Journal of the History of Sexuality* 14 (January 2005/April 2005): 29.
2. Monica J. Casper and Laura M. Carpenter, "Global Intimacies: Innovating the HPV Vaccine for Women's Health," *Women's Studies Quarterly* 37 (Spring/Summer 2009): 80–100.
3. Lisa Jean Moore and Mary Kosut, *The Body Reader: Essential Social and Cultural Readings* (New York: New York University Press, 2010).
4. Steve Seidman, Nancy Fischer, and Chet Meeks, *Introducing the New Sexuality Studies: Original Essays and Interviews* (New York: Routledge, 2007), xii.
5. Michael S. Kimmel and Rebecca F. Plante, *Sexualities: Identities, Behaviors, and Society* (New York: Oxford University Press, 2004), 4.
6. Janice Irvine, *Disorders of Desire: Sex and Gender in Modern American Sexology* (Philadelphia, Temple University Press, 1990), 5.
7. Ibid., 83.
8. Gayle Rubin, "Thinking Sex: Notes for a Radical Theory of the Politics of Sexuality, " in *The Lesbian and Gay Reader*, eds. Henry Abelone, Michele Aina Banale, and David Halperin (New York: Routledge, 1993): 11.
9. Ibid., 9.
10. Ibid., 16.
11. Monica J. Casper and Lisa Jean Moore, *Missing Bodies: The Politics of Visibility* (New York: New York University Press, 2009), 23.
12. For example, although not specifically and exclusively dedicated to sexuality studies, the *Journal of the American Medical Association*, the *Journal of the American Public Health Association*, and the *New England Journal of Medicine* (among others) do publish relevant articles about human sexuality, but we did not include these journals because we do not consider them central to the field of new sexuality studies—although indeed their biomedical scholarship does serve as data for sexuality studies scholars because they frame public health, surveillance and state administered human services related to sexuality. The web-based community of critical sexuality scholars has performed their own content analysis available at http://sexualitystudies.net/content/short-background-critical-sexuality-studies. Researchers examine and address sexuality studies as it appears in articles, journals, interviews, and consultations between the years 2000 and 2006.
13. Chris Beasley, *Gender & Sexuality: Critical Theories, Critical Thinkers* (Los Angeles: Sage, 2008), 118.

14. Susan Stryker and Stephen Whittle, *The Transgender Studies Reader* (New York: Routledge, 2006), xii.

15. Heather Love, "'Oh, the fun we'll have': Remembering the Prospects for Sexuality Studies," *GLQ: A Journal of Lesbian and Gay Studies* 10 (2004): 259.

16. Dean Spade and Sel Wahng, "Transecting the Academy," *GLQ: A Journal of Lesbian and Gay Studies* 10 (2004): 248.

17. Patricia Elliot, "Engaging Trans Debates on Gender Variance: A Feminist Analysis," *Sexualities* 12 (February 2009): 5.

18. Ibid., 11.

19. Jeff Hearn, "Sexualities Future, Present, Past... Towards Transsectionalities" *Sexualities* 11 (February 2008): 39.

20. Arlene Stein, "From Gender to Sexuality and Back Again: Notes on the Politics of Sexual Knowledge," *GLQ: A Journal of Lesbian and Gay Studies* 10 (2004): 256.

CHAPTER 8

"THE BUGS OF THE EARTH": REFLECTIONS ON NATURE, POWER, AND THE LABORING BODY

DIANA MINCYTE

BY AND LARGE THERE IS NOT ENOUGH BODY IN ENVIRONMENTAL STUDIES, as practiced in North American academic traditions. This may be because it falls through the cracks between two main paradigms defining the field. The first paradigm reaches back to the early decades of the twentieth century when the powerful minds of the time (including Gifford Pinchot and John Muir) called to consider anthropogenic—and deeply destructive—impacts on the environment and pushed for concerted efforts to conserve, preserve, and protect nature. Although the emerging field of environmental studies proposed moving beyond human-centric treatment of the environment, their approaches, when put into practice, excluded nonwhite bodies, displacing entire populations of inhabitants, erasing their social memory, and reconfirming the dominance of the modern state and its institutions over territories. In today's scholarship, such approaches are often critiqued as "fortress preservation," yet they still surface in the debates about biodiversity and sustainability, albeit in more conceptually elaborate and self-reflexive forms.

The second paradigmatic shift took place more recently with a revival of power-inflected understandings of the interface between the social and the biological. Culminating in William Cronon's edited volume, *Uncommon*

Ground: Rethinking the Human Place in Nature (1996), these approaches call for a more situated perspective to examine how social, political, and economic forces play out in human relations to nature.[1] Following this trend, Barbara Bender, for example, argues that people's participation in nature "is based in large measure on the particularity of the social, political and economic relations within which they live out their lives."[2] Although these approaches bring the political, the economic, and the social to bear on the material, they often lose sight of the bodies that are not simply mediators of social and political relations, but also act as material agents plugged into ecological systems.[3]

For obvious reasons, this highly stylized description of the field does not do justice to a number of scholars who have sought to capture the body as being in and partaking of the environment and, by so doing, questioning where the body ends and nature begins. Some telling examples come from debates about technosciences, posthumanism, multinaturalism, and ontological shifts as well as from the fields of ecofeminism, science studies, and political ecology. But these studies, while challenging the field, have been few and far between, and they have generally been pushed outside of the margins of environmental studies textbooks. Such marginalization seems to be rooted deeply in the Enlightenment's boundary-making projects producing a corpus of literature where messy subjects and relational experiences do not comfortably fit inside the disciplinary grid.

In this chapter, I reflect on what it would mean for the body to be part of the "natural" landscape and, more broadly, an agent in the field of environmental studies. Working within the parameters of practice-based, power-inflected, and subjectivity-driven approaches that were articulated in the work of Maurice Merleau-Pointy, Arturo Escobar, Timothy Ingold, and Judith Butler, I hope to sketch the contours of the body-in-nature approach. The body in this approach is not limited to the confines of skin, but extends—through practices and power relations—to material landscapes and political domains. This approach is deeply indebted to Judith Butler's emphasis on relationality, but rather than focusing on the human body, I seek to reflect on how social power is practiced in and through bodies in nature.[4]

Apart from being an exercise in conceptual thinking, this project draws on my fieldwork and archival research on agricultural labor in Soviet and post-Soviet Lithuania to explore how such an approach plays out in a specific case. In this sense, the chapter is set up to show more and tell less by giving detailed and thick descriptions of practices, landscapes, and subject positions, and leaving brief reflections about the implications of environmental biopolitics for the concluding section. The case presented here explores how rural populations in the post-World War II Soviet Union experienced nature, a question about relational connections among local landscapes, rural subjectivities, and the noncapitalist, authoritarian state. Entangled deeply in every aspect of this triangle, the laboring body emerges as the key component in defining and enacting connections between the land, the subject, and the Soviet state. Yet, what is revealed by this case is not only that the body figures centrally in biopolitics and landscaping practices, but also that the notion of the environment itself needs to be broadened, both deepened and opened up to match the relational, political, and material dimensions of the body. Ultimately, this chapter calls for a more animated and nuanced understanding of the biopolitics of the environment.

THE POLITICS OF REPRESENTATION

From the very first day of my fieldwork in Lithuania, I faced two major challenges. The primary issue was related to the overly politicized subject of my research in which the issues of landscapes and local ecologies tended to recede to the backdrop for larger political drama. Volumes had been written on the Baltic States in the throes of the postwar Soviet occupation, local ideological warfare, ongoing civil unrest, collectivization, exiles, torture, and persistent hunger (e.g., in the work by Romuald Misiunas, Rein Taagepera, and Liudas Truska). The body of the Soviet farmer was very much present in this scholarship, but it was suspended in the political webs spun by the Soviet state and disconnected from the environments and life on the farm.[5] From oral histories, participant observations, and archival documents, a different story emerged where food, labor, and land were central in the daily lives of Soviet farmers struggling to feed themselves and their kin. In other words, rather than openly engaging in ideological battles, the daily lives of the Soviet farmers living on the verge of famine seemed to have revolved around subsistence and work. Following this line, a methodological question of how to rethink the suffering body in the context of daily practices and local landscapes became central for this project.

Second and relatedly, at the heart of this project was the question of agency. In the opening pages of his book *The Mental and the Material: Thought, Economy and Society* (1996), Maurice Goddelier states that people have history because they transform nature, suggesting that human agency is constituted at the interface between the natural and the social.[6] From this perspective, the Soviet farmers' ability to transform nature into food made them into important historical agents. And yet, the voices and daily practices of the laboring farmer were missing in the pages of postwar histories of the Baltic States. Portrayed as victims of the regime, the farmers in aggregate were considered as external to the processes of rapid modernization, industrialization, and urbanization taking place in the Soviet Union. Although they were not represented in the history, their bodies and their labor on the land made the very existence of the Soviet state possible. In such a manner, by following Goddelier's conceptualization of historical agency, I ask, in what ways did this embodied agency of the peasantry come into being? Also, how did their daily practices translate into political agency?

In what follows, I present a story about the making of corporeal agency and political subjectivities in the Soviet Union through agricultural labor. By looking at the Soviet state and how collective farm members worked their small land plots and how they transformed Soviet landscapes and nature through their labor, this chapter draws connections between the embodied experiences of nature and the construction of state power under socialism.

THE HISTORY OF SUBSIDIARY FARMING

The history of subsidiary farms extends back to the early postrevolutionary period in Russia when requisitions were imposed on farmers to ensure food supplies for the struggling cities. Even though the leadership of the Communist party was suspicious of peasantry as supporters of private property and thus potential ideological enemies, the young state was also painfully aware of the significance of food supplies. Early in the 1920s, the party made major concessions to the farmers by allowing them to keep part of their land and even to hire private labor. These reforms were soon discredited as counterrevolutionary and by 1929 abolished, but they proved to be significant in that they established a precedent in Soviet history regarding the strategic use of peasants' skills, knowledge, and bodies for fueling industrialization.

The first subsidiary farms were officially founded in 1933 following the 1930 Model Charter permitting the issuance of small-scale allotments along with pasture plots for use by collective farm members and state farm employees. The size of the subsidiary farms was limited to 50 ares (1.24 acres), while pastures were capped at 1 hectare.[7] In addition to land, farmers were also allowed (and in some cases, forced) to keep livestock: one cow (two cows in Lithuania), two calves, one sow with young or two sows with young in cases where collective farm authorities decided it was necessary, ten sheep and/or goats, an unlimited amount of poultry and rabbits and up to twenty beehives. During Khrushchev's rule, direct requisitions from these diverse farms were replaced with a contract system where the farmers signed agreements committing themselves to grow the livestock and sell it to the collective farm for a preagreed price.

When the new Soviet republics—Latvia, Lithuania, and Estonia—were incorporated into the Soviet Union in 1944, the size of the subsidiary farms was increased to 60 ares.[8] In Lithuania, subsidiary lots were usually divided into two parts. The first part was a kitchen garden of about 15 ares located in close proximity to the house. The location of the second, larger plot was determined by the collective or state farm leadership, and it could change from year to year depending on leaders' often capricious decisions. The farmers could grow almost anything on their subsidiary farms, and the only important exception was a ban on growing the same crops as those in collective farm fields, especially fodder beets or grains. As several former state farm employees interpreted, these restrictions were supposed to make it easier for the supervisors to prevent stealing. Not surprisingly, in the context of sweeping famine, ongoing requisitions, and absence of any other sources of food or cash incomes, these subsidiary lots were enormously productive. Occupying less than 7 percent of all agricultural land in the Soviet Union, these farms constituted more than half of all key agricultural production well into the 1960s.[9]

Despite the fact that the Soviet farms never had full-fledged rights to private ownership of the land, there is a tendency in the scholarship to call subsidiary farms "private."[10] This is primarily because the profits from farming went to individuals/households. And of course, it was individual household members (and their animals) who did the work. In practice, however, the Russian *pri-usadebnyje uchastki* (farmsteads or estates) and Lithuanian *sodybiniai sklypai* (garden plots) refer not so much to ownership rights, as to the territory or the material space that these plots occupied in farmers' lives. More specifically, the terms refer to the proximity of the land plot to the residence of the farmer. In Lithuanian, the term *sodyba* (farmstead) invokes the farmstead that included the land, the built structures, gardens, as well as human beings. Although economic incentives were extremely important when working on the plots, the land on which farmsteads stood was imbued with multiple symbolic and cultural meanings that went far beyond economic interests or ownership rights. Fruit trees, vegetable garden, bee hives, flowers, the house, the barn, the well, the swings, the pond, cows, pigs, cats, dogs, frogs, birds, snakes, and insects—all these elements in the *sodyba* weaved not only into the everyday practices, but also into the mythologies of the place. Children, as oral history reveals, knew all too well about different types of deities especially those inhabiting the barn, the well, and the threshold, and acted accordingly. The human life transitions—birth, marriage, and death—were all performed in this place and belonged to its natural cycles. On these multiple levels, the body of the peasant was plugged into the lifecycles of *sodyba*.

THE MAKING OF LANDSCAPES AND SUBJECTS

Since kitchen gardens were located near homes, the farmers knew the soil and its drainage quite well, and they planted vegetables, fruit trees, and herbs and established bee hives in the areas where they knew the conditions to be most appropriate. In the process of carefully arranging and reconfiguring the garden, they created patchwork plots that followed the contours of varying material qualities of land. In lower areas and where the soil was not draining well, they planted bushes. Trees were usually grown close to the house or at the north end of the plot so as to leave the most sunlight for the main garden as well as to protect it from the wind. Beans, peas, and most herbs grew in the shade of the trees, while carrots, beets, and onions were planted in well-drained areas. Where the soil was clayish, cabbages were planted in long rows interspersed with garlic, calendula, chamomile, dill, and other herbs. Many farmers kept bees, and in some cases, neighbors had bee hives next to each other allocating a shared space for them.

Such an organization of the garden embodied a complex relationship between the farmer, the land, and the surrounding environment where attempts to carefully control the garden intersected with an acknowledgment of the limits of human agency. On the one hand, one's long-term embodied engagement with the garden enabled the farmers to impose a certain cognitive map or an order onto the landscape. They planted vegetables exactly where they thought they could grow best and where they knew the soil was most suitable for the current year.

In terms of human subjectivities, such practices require one's immediate and long-term engagement with the material agency of the garden. It is a commitment to observe the transformations, fight external powers, interact with soil, worms, birds, vegetables, weeds, the sun and water, and, most importantly, put in long hours of manual work. Such practices make the farmer an inseparable part of the ecological assemblage of the garden, but it also made them the masters of their domain. The concept of a "good farmer" as a moral category surfaced in many interviews suggesting a significant intersection between practices of gardening and the local social order. The "good farmer" was the one who kept her or his garden orderly and weed-free. The best farmers also knew how to protect seeds, weather bad years, prepare the soil, deal with animals, and maintain social relations with the neighbors.

On the other end of the spectrum, the farmers saw themselves as powerless in the face of nature's forces. In every interview and observation, they acknowledged that they fought a losing battle to keep the gardens orderly and productive. As an older woman who used to be a milkmaid in a collective farm pointedly stated, "Garden is not a machine. You cannot turn it on and off when you want it and where you want it. Sometimes you get rewarded for slacking and not planting on time, and at other times, only the hardest work pays off."[11]

The recognition of farmers' limited ability to control the environment and the unpredictable outcomes of work is best captured in the metaphor of the "bug of the earth" (Lith. *žemės vabalėlis*) that was referred to in several interviews. In using the diminutive form of the word "bug" (*vabalas* vs. diminutive *vabalėlis*), the farmers emphasized the small size of the "bug," but also its submission to the powers of nature. The metaphor of the *bug of the earth* captures the humble role that the farmers saw themselves playing in the agricultural ecology. Through this language they emphasized that they were fully dependent on the garden and were feeding on the garden's fruit, and that their efforts to work for the garden were minute in comparison to those of nature.

In the light of these observations, gardening practices involved a careful fine-tuning of the environment, human knowledge, and vegetables. To capture the farmers' attempts to steer the growth processes, it is helpful to use Andrew Pickering's (1995) notion of the dance of agency.[12] Pickering's dance of agency relates to the continuous interaction between science and its subjects. Just like the scientists who go back to experiments in response to the challenges posted by materials they study, so too Lithuanian farmers dealt with the human and nonhuman agencies to bring the garden into the desired state. Weeding, watering, fertilizing, and dealing with pests are precisely the ways of "dancing" that allowed the farmers to keep the garden approximately within the certain desired parameters and yet be profoundly affected and open to the unpredictable forces of nature.

LANDSCAPING AS BIOPOLITICS

World War II left East Europe devastated. But even war devastation did not compare to the effects of nationalization and collectivization that began soon after the war in the three Baltic States. The Soviet state did promise to let the farmers keep a plough, a horse, a cow, two sheep, and seeds for the coming season as well as a line of clothing—a suit, a coat, two shirts, a hat, one pair of shoes, and one fur coat for the whole family. But these promises were rarely kept. Reports on the nationalization of households at the local archives testify to the brutality when all furniture, linens, clothes, dishes, or utensils were taken away, not to speak of the animals, tools, and seed. It is in this context of scarcity, fear, and famine that small plots of land became the lifeline for everybody and the epicenter of biopolitics in rural societies.

Instead of using "naked power" to push the farmers to work on collective farm fields, the collective and state farm leadership knew very well that the location of the subsidiary farm and pastures was the most effective string in controlling the farmers' bodies and behavior. They could reward their allies and "good Soviet citizens" with better land in close proximity to their homes or condemn one to laboring on stone-filled, weed-ruled, infertile clayish soils that were located miles away from their homes. Unlike supervision technologies that produced self-disciplined subjects of the state in Western European and North American contexts, such place-based and environmentally mediated forms of power worked through the relationship of the human and the land and through the leverage of subsistence.

The emphasis on the practice and the environment as methodologies of power is deeply embedded in the larger biopolitics of the Soviet state. In his work on Stalin's biopolitics, Amir Weiner argues that there are fundamental differences between the Nazi ideology of biological determinism and the Soviet ideology of purging that exercised biopower. Building on Zygmund Bauman, Peter Holquist, Terry Martin, and James Scott, Weiner develops a power metaphor of a "gardening state."[13] In employing this metaphor, Weiner argues that after long years of class warfare followed by World War II, Stalin's government attempted to construct a harmonious and peaceful "garden-state" and to do so, the Soviet state used violence to weed the contaminated, polluted, and parasitic elements from its social body.

In a different way than Nazism, however, the Soviet leadership was concerned less with physical/biological rather than social/cultural origins of "parasitism." This approach meant that specific ethnic groups would not be targeted for full extermination as was the case in Nazism, but the side effects of such a method of governance also included an omnipresent suspicion over who was who: anybody, anywhere and at any point of time could turn into an enemy. In this sense, the target of the Soviet "gardening" practices was not the body per se with its specific genetic composition, but the minds and hearts of its citizenry and the "malleability" of human nature, more broadly.

Most importantly, the locus of state's power was the relationship of the body to the social, political, and material environments. By moving away from the body as the source of society's "pollution" to focusing on the actions of its citizens put increased pressure on everyone to publicly perform their ideological alliances and support for the Soviet state.[14] For the farmers, agricultural labor on subsidiary farms and especially fulfilling requisitions were precisely the methods for signaling their political innocence. In response to extraction and extermination, the farmers turned to working the land harder and by fulfilling production quotas through deliveries of produce and livestock to the state, they performed proof of their ideological integrity. As a result, their laboring bodies and the food they produced on the land became their ticket to the future, while feeding and supporting the state. In this complex interface among violent "gardening" state, landscape politics, labor, and the bodies of the farmers, the work on the subsidiary farms and the values of "good farmer" became aligned and integrated into the Soviet state building projects.

*   *   *

There are at least two implications that follow from the land and labor based biopolitics that I outlined through the case of subsidiary farming under socialism. First, current debates on the body and biopolitics tend to focus on the advancement of biotechnologies and are concerned with the growing incisions that technosciences, neoliberal ideologies, expert regimes, and regulatory institutions make to our bodies, reaching even their molecular processes. In this context, my project suggests that we also consider spatial and environmental methods as sites for exercising biopower. While these subsistence and physical labor driven politics of the body may seem far removed from the deepening immunity politics, the reconfiguration of governance institutions, or reconceptualization of sovereignty as outlined in the work of Nikolas Rose, Roberto Esposito, and Michael Hardt and Antonio Negri, they remind us of the more "raw" forms of governance that coexist side by side with these internal and identity-centered biopower projects. In this sense, a move to rethink geographies and environments through and on which biopolitical projects are implemented may allow inclusion of more diverse perspectives as well as social groups who inhabit (physically and socially) the margins of postindustrial, consumer societies.

For environmental studies, integrating the body into the classical questions that define the field such as the politics of land-use and water rights, environmental justice, nature preservation, nonhuman agency, global climate change, and others, this project underscores the need to develop more nuanced approaches to capture the deeply contradictory and complex issues surrounding nature's agency and its place in social relations. When we juxtapose the most recent theories of the body that conceptualize it as a relational entity constituted through social practices, scientific institutions, technologies, and materiality, with a notion of the environment that is still often confined within the debates about what is natural and what is human, it becomes clear that questions of governance of the environment and its biopolitics are still up for grabs.

NOTES

I would like to thank Monica J. Casper, Paisley Currah, Zsuzsa Gille, Andrew Pickering, Neringa Klumbyte, and the anonymous reviewer for their insightful comments on the earlier versions of this chapter. I am also grateful to the villagers in Lithuania who shared their life stories with me, making this research possible.

1. William Cronon, ed., *Uncommon Ground: Rethinking the Human Place in Nature* (New York: W.W. Norton, 1996).

2. Barbara Bender, "Stoneage—Contested Landscapes," in *Landscape: Politics and Perspectives*, ed. Barbara Bender (Oxford: Berg), 245–279, 246.

3. Marina Fischer-Kowalski and Helga Weisz, "Society as Hybrid between Material and Symbolic Realms: Toward a Theoretical Framework of Society-Nature Interaction," *Advances in Human Ecology* 8: 25; Bruno Latour, *Politics of Nature: How to Bring the Sciences into Democracy* (Cambridge, MA: Harvard University Press, 2004); Timothy Ingold, *The Perception of the Environment: Essays on the Livelihood, Dwelling, and Skill* (London: Routldege, 2000).

4. By biopolitics here, I refer to the work of Michel Foucault and especially his analysis of biopower and the methods through which government regulates populations. Following Monica Casper, Katherine Hayles, Roberto Esposito, Jacques Derrida, among others, I approach this concept as the locus of technological advancements, public policies, embodied experiences, and social relations that are reconfigured to exert new ways of controlling various forms of human and nonhuman life.

5. For more on the issues surrounding the body and rural societies during collectivization in the Soviet Union, see Lynne Viola, V. P. Danilov, N. A. Ivnitskii, and Denis Kozlov, eds., *The War against the Peasantry, 1927–1930: The Tragedy of the Soviet Countryside*, trans. Steven Shabad (New Haven, CT: Yale University Press, 2005); Lynne Viola, "The Last and Most Decisive Battle: Collectivization as Civil War," *Peasant Rebels under Stalin: Collectivization and the Culture of Peasant Resistance* (Oxford: Oxford University Press, 1996), 13–44; Robert W. Davies, *The Socialist Offensive: The Collectivisation of Soviet Agriculture, 1929–1930* (Cambridge, MA: Harvard University Press, 1980); Robert Conquest, *The Harvest of Sorrow: Soviet Collectivization and the Terror-Famine* (New York: Oxford University Press, 1986); Sheila Fitzpatrick, "New Perspectives on Stalinism," *Russian Review* 45, no. 4 (October 1986): 352; Sheila Fitzpatrick, "Introduction," in *Stalinism: New Directions*, ed. Sheila Fitzpatrick (London: Routledge, 2000), 1–14; Stephen Kotkin, *Magnetic Mountain: Stalinism as a Civilization* (Berkeley: University of California Press, 1995).

6. Maurice Goddelier, *The Mental and the Material: Thought Economy and Society*, trans. Martin Thom (London: Verso, 1996).

7. An *are* is a metric unit of area equal to 100 square meters.

8. "Dėl Kolūkių Organizavimo Lietuvos TSR" ["About Collective Farm Establishment in Soviet Lithuania"], Lietuvos TSR Aukščiausiosios Tarybos ir Ministrų Tarybos Žinios [The News Bulletin of the Supreme Soviet of the Soviet Republic of Lithuania] 10, no. 56 (1948): 12.

9. Karl-Eugen Wadekin, "Soviet Rural Society," *Soviet Studies* 22 (December 1971): 512; Karl-Eugen Wadekin, *The Private Sector in Soviet Agriculture*,

trans. Keith Bush, ed. George Karcz (Berkeley: University of California Press, 1973); Moshe Lewin, *Russian Peasants and Social Power*, trans. Irene Nove (London: George Allen and Unwin, 1968); Gelii I. Shmelev, *Lichnoe Podsobnoe Khoziaistvo: Vozmozhnosti i Perspektivy* (Moscow: Mysl', 1983); Gelii I. Shmelev, *Personal Subsidiary Farming under Socialism* (Moscow: Progress, 1986).

10. F. Khiliuk, "Lichnoe Podsobnoe Khoziaistvo Naseleniia i ego Rol' v Proizdvostve Sel'skohoziaistvennykh Produktov" ["Individual Subsidiary Farms and Their Role in the Agricultural Production]. *Ekonomika Sovetskoi Ukrainy* 1, 1966; Wadekin, "Soviet Rural Society"; Stefan Hedlund, *Private Agriculture in the Soviet Union* (London: Routldege, 1989).

11. Interview conducted on July 17, 2004.

12. Pickering, Andrew, *The Mangle of Practice: Time, Agency, and Science* (Chicago: University of Chicago Press, 1995).

13. Amir Weiner, *Making Sense of War: The Second World War and the Fate of the Bolshevik Revolution* (Princeton, NJ: Princeton University Press, 2001); Amir Weiner, "Nature, Nurture, and Memory in Socialist Utopia: Delineating the Soviet Socio-Ethnic Body in the Age of Socialism," *American Historical Review* 104, no. 4 (June 1999): 1114–55; and Amir Weiner, ed., *Landscaping the Human Garden: 20th Century Population Management in a Comparative Framework* (Palo Alto: Stanford University Press, 2003).

14. Alexei Yurchak, *Everything Was Forever, Until It Was No More* (Princeton, NJ: Princeton University Press, 2006); Alexei Yurchak, "Soviet Hegemony of Form: Everything Was For Ever Until It Was No More," *Comparative Studies in Society and History* 45, no. 3 (2003): 480.

CHAPTER 9

THE AUDIBLE BODY: RFIDs, SURVEILLANCE, AND BODILY SCRUTINY

SHOSHANA MAGNET

IN A SEGMENT CALLED "FUTURE SHOCK," Jon Stewart's nightly comedy news show team interviewed technophile Mikey Sklar, one of the first American citizens to elect to have a subcutaneous radio frequency identification device (RFID) injected into his arm. When questioned by a Daily Show reporter as to the reason that he chose to be implanted with the chip, Sklar reported that this innovative technology allowed him to accomplish many new tasks:

> *Samantha Bee: What can you do with this chip?*
> *Sklar: Well, I can hold my hand in front of a deadbolt, and the deadbolt will throw.*
> *Samantha Bee: Wait a sec, whoa. You can open a door... with your hand? Well, I'll believe that when I see it.*[1]

Problematizing the hype around new technologies used to accomplish old tasks, the above humorous example introduces my investigation of subcutaneous RFID chips. In this chapter, I unpack the relationships among surveillance, technology, and the human body. Since 9/11, we have witnessed the rise of a multitude of new identification technologies that emphasize the productive potential of the human body for the governance of the state. From biometrics to backscatter x-rays, new ways of scanning the body in the name of security abound. Individuals are increasingly linked through their bodies into networks as part of what Simone Browne calls the "identity-industrial complex," in which the body itself becomes the primary object of surveillance.[2] Many of these technologies involve new ways of visualizing the body. Backscatter cameras render the body virtually nude, infrared technologies make bodily heat visible, and biometric technologies transform the body into binary code.

Whether it is screening women's bodies at security checkpoints or policing immigrants at the US-Mexico border, practices of looking at the human body remain tied to assumptions around gender, race, class, sexuality, and disability. Although renewed efforts to visualize the body represent an important trend within contemporary forms of both state and corporate surveillance,[3] in this chapter I examine another post–9/11 phenomenon: the development of radio tag technologies that may be implanted subcutaneously to compel the body to speak. This is a process that I term "making the body audible." RFIDs are tiny wireless chips digitally encoded with identifying information. Previously, RFIDs primarily were used in consumer applications such as inventory control as well as in livestock and companion animal tracking. Recently, the US Food and Drug Administration (FDA) approved an implantable RFID for use in humans. This chip, about the size of a grain of rice, makes the United States the first nation in the world to allow digital chips to be inserted into the human body.

Using critical communication theory to understand the ways that RFIDs are being used to transform understandings of bodily interiority, I argue that contemporary state practices of surveillance are emphasizing communication as a new way of compelling the body to speak its identity. Highlighting the ocular-centric tendencies of surveillance studies as a field, I interrogate the connections between surveillance and sensory practices more broadly. Moving beyond "line-of-sight" technologies that require direct visual contact such as the backscatter x-rays and other visual technologies mentioned earlier, the major contribution of this chapter is that communication is central to thinking about surveillance and the body. Although RFID chips are communication technologies, dependent on radio spectrum and regulated by the FCC, the ability of implanted chips to make bodies audible has not yet been addressed. Here, I document the ways that RFID chips are represented as able to compel formerly silent bodies to speak—enabling us to theorize the sounds of surveillance.

In addition to demonstrating how bodies are made audible through contemporary forms of surveillance, I also interrogate how new technologies are used to construct bodies as forms of human inventory. RFID discourse manufactures bodies as newly locatable as chips are inserted in order to find individuals in both time and space—a key feature of any database technology. Whereas the panopticon renders bodies two-dimensional, I show that RFIDs are deployed to penetrate inner bodily space, producing the body under surveillance in three dimensions. Although RFID technologies are hyped to be able to accomplish the new task of sorting people with "mechanical objectivity,"[4] like Mikey Sklar who uses an RFID chip to complete a task he could already do, I show that RFIDs are being used for the familiar task of categorizing people in ways that reinforce existing systems of inequality. Instead of giving formerly silenced bodies a place at the conversational table in important national discussions about issues such as poverty, racism, or immigration, we will see that attempts to implant RFID chips in certain communities fill vulnerable bodies with a very particular type of speech in the hopes of furthering corporate profits. Thus, rather than freeing othered communities from discrimination, RFID discourse attempts to turn silent subjects into communicating objects, yielding new kinds of bodily truths about social location and national identity, including race, ethnicity, class, gender, disability, and sexuality. Of course, subcutaneous RFIDs are not applied without a struggle. In examining some of the forms of resistance to subcutaneous RFID chips, I show that, rather than simply being used to subjugate bodies, instead RFIDs provide a useful window into contested conversations about state-making in the age of security.

I begin with a brief history of the intersection of state and corporate applications of RFID technologies to human bodies, as corporations seek to secure the state as a client for their technologies to substantially increase profits. I then describe my theoretical framework in order to investigate the construction and regulation of RFIDs as communication technologies. Through an analysis of the early places that RFIDs were proposed, including the attempted application of implanted RFIDs to immigrants and refugees to the United States, this chapter aims to intervene in the narrative of uncomplicated technological progress articulated to promote the adoption of this new technology. In foregrounding that RFIDs remain a field of struggle and contestation between those in power and vulnerable groups, mediated by the governmentalities of security, we see that while RFID technologies make the body newly audible, so too do they amplify the politics of surveillance.

FROM MERCHANDISE TO HUMAN INVENTORY

> That same scanner in a Wal-Mart that is used to bar-code your goods can be used to identify you...
> —Scott Silverman, CEO of Applied Digital Solutions[5]

The first use of RFIDs for human tracking was in 1994, when some 50,000 refugees fleeing violence, poverty, and political repression in Cuba and Haiti were captured by the United States Coast Guard and held under deplorable conditions in Guantanamo Bay, Cuba.[6] Upon their arrival, adults were issued RFID wristwatches (children had to wear RFID anklets), which they were not allowed to remove. Since then, external RFID chips meant for human tracking have been used in multiple locations. Schools in both the United States[7] and Japan[8] have required children to wear them. In 2005, 1,800 prisoners in Castaic, California, were issued RFID wristwatches in a pilot project costing $1.5 million.[9] By 2008, prison inmates at correctional facilities in Virginia, Michigan, Illinois, Ohio, and Minnesota were wearing RFID tags.[10] Although the use of RFID wristwatches and other tags on the most vulnerable segments of the population—whether children, refugees, or prisoners—is troubling and begs further analysis, here I investigate the deployment of RFID tags implanted under the skin.

In October 2004, the US FDA approved a patent allowing RFID chips to be implanted subcutaneously.[11] Verichip[12] is the company responsible for the subdermal tags, which it markets as a fail-safe form of identity verification: "Unlike traditional forms of identification, the Verichip can't be lost, stolen, misplaced, or counterfeited. Because it's inserted under the skin, it's always there when you need it regardless of where other kinds of identification might be."[13] Using an inserter as large in diameter as a knitting needle in what Verichip™ somewhat dubiously claims is a "quick, painless outpatient procedure,"[14] the chip is implanted into subcutaneous tissue. The chip is encoded with a sixteen-digit identification number, which, when scanned, allows medical personnel to access the person's medical records if they have been uploaded and linked to the person's chip. This ostensibly allows hospitals with the appropriate scanning technology to quickly access information from an unconscious patient or a patient having difficulty communicating. Verichip™ highlights patients with Alzheimer's or suffering from dementia as prime candidates for the technology.

How does RFID technology work? A way of automating identification, radio, or RFID tags allow for information to be remotely stored and retrieved. Usually, an RFID tag contains two components, the circuit and the antenna.[15] The circuit (a small silicon chip) is what allows information to be stored and retrieved, and both circuit and antenna are stored within a small tube of unbreakable glass. Implantable chips are commonly described as about the size of a grain of rice, although chips may be much smaller. The RFID reader emits radio waves constantly, waiting to see if there are any RFID tags in the area.[16] When an RFID tag gets close enough to a reader, the RFID chip's antenna picks up the radio waves being emitted by the reader and amplifies the reader's energy.[17] The chip then exchanges information with the reader. Of course, readers may be linked into databases that in turn may be linked into networks, allowing additional information to be accessed.

An active radio tag contains its own source of power, most often a battery.[18] A passive RFID must communicate remotely with a transponder in order to have power. Implantable RFID chips are currently passive, since it would be far too difficult to surgically remove them in order to change their batteries. Passive tags have much shorter "read" ranges than active tags.[19] That is, one must be closer to a tag reader for a passive tag than an active tag to work. Moreover, an active tag is able to actively communicate the information encoded on it rather than waiting passively to come into a contact with a reader to function.

SURVEILLANCE BEYOND "LINE OF SIGHT" TECHNOLOGIES: COMMUNICATING THE INTERIOR

Many security technologies take the body itself as the object of surveillance. Backscatter x-rays strip the body naked, closed circuit television (CCTV) records bodily movements in space, whereas biometric technologies render human bodies as binary code. Of course, contemporary security technologies are the latest development in a long history of technologies used to visualize the body, from portrait painting and photography to early systems of criminal identification such as bertillonage and fingerprinting, compelling the body to speak the truth of its identity is a thoroughly modern tale. That is, uses of these technologies make subjects modern in regimes of surveillance. Rachel Hall suggests the phrase "aesthetics of transparency" to describe post–9/11 preoccupations with security.[20] Demonstrating the ways that state and corporate institutions imagine perfect security to depend on unimpeded visibility, Hall shows that contemporary security technologies claim to be able to see *through* opaque bodies into their hidden corners to ascertain whether they conceal a threat.

Hall's research draws upon earlier work done by feminist science studies theorists, including Lisa Cartwright, Donna Haraway, and Paula Treichler, all of whom problematize the notion of perfect, objective vision. They instead point to ways that new visualization technologies used to screen the body, rather than rendering the body transparent, tend to "reinforce what we have already learned to see."[21] In scrutinizing contemporary preoccupations with the transparent body as a necessary precursor to perfect security, we begin to see how the body under surveillance is one that state and corporate institutions attempt to make see-able, code-able, and know-able in order to facilitate its control.[22] I turn now to the ways that RFIDs are part of attempts to make the body audible by compelling the body to speak.

Like other communication technologies, radio tags are regulated by the Federal Communications Commission (FCC) in the United States.[23] RFID technologies, like any other radio wave technology, depend on frequencies and thus must be allocated spectrum. Radio tags are currently allocated a piece of the spectrum referred to as UHF, or ultra high frequency, which is found between 860 and 960 MHz. The RFID industry regularly describes the utility of radio tag technology in terms of its ability to produce new forms of communication. For example, one of the metaphors used to describe the proliferation of RFIDs is the Internet of Things.[24] The Internet of Things is not a new communication network. Rather, grafted onto the top of the Internet, what is significant about the Internet of Things is that it would give every object its own, unique identifying tag. As a result, every object, whether animate or inanimate, "would be endowed with the ability to talk to manufacturers, retailers, and even each other."[25] Of course, this means that penetrating individuals with these new communication technologies links bodies into communication networks such that they can be compelled to speak with (and on behalf of) state and corporate institutions in new ways.

Industry rhetoric describing the potential of radio tag technologies is rife with communication metaphors. As Verichip™ says, its chip intended for medical purposes, named VeriMed, "is always ready to 'speak' on your behalf by quickly providing healthcare professionals with your name and pertinent medical information."[26] In a second example, in describing the benefits of their implanted RFID technologies for patients, Verichip™ emphasized its ability to endow silent bodies with the power of speech and to perfect poor communication between doctors and patients. That is, if someone has difficulty communicating their health concerns—or if a person is unconscious—the suggestion is that Verichip™ will be able to step in to fill in the communicative void. As the company states on its homepage, "By 'speaking' for patients with chronic illnesses, VeriMed offers an empowering option to affected individuals. The application helps these types of at-risk individuals to obtain a comparable level of care by rapidly and accurately furnishing important or even lifesaving information when and if they are unable to do so."[27] Equating communication with empowerment, implanted RFID tags are imagined to be able to productively fill bodily silence. These examples also reveal the ways that RFID discourse produces bodies as human inventory.[28] A key feature of any inventory technology is an improved ability to locate the object in time and space, and RFID chips make it easier to locate unconscious bodies both temporally and spatially. In including these chips in patients, we see how the RFID industry aims to mine unconscious bodies for profit, as bodies that are rushed into the medical ward unable to speak are compelled to communicate the information included on their implanted chips.

Earlier, I historicized the development of subcutaneous RFID technologies to show that regimes of surveillance and their associated technologies do not end with making the body visible, but remain interested in making the body audible. That is, the body placed under surveillance is productive not only of new forms of visual information, but also of audio information. In examining the production of new bodily sounds that connect individuals into modern networks of surveillance, we can understand how new surveillance technologies compel bodies to produce not only "sightbytes,"[29] but also soundbytes. A reliance on communication metaphors is characteristic of contemporary scientific narratives. Paraphrasing Haraway, Paula Treichler observes that modern immunology could only move "into the realm of high science when it reworked the military combat metaphors of WWII (battle, struggle, territory, enemy, truce) into the language of postmodern warfare: communication command control (coding, transmission, messages), interceptions, spies, lies."[30] Similarly, reworking the body under surveillance in terms of communication metaphors situates RFID technologies firmly within the present moment. The RFID industry's suggestion that there exists a technology that can achieve the dream of perfect communication is not new. From the telegraph to the Internet, we are constantly told that communication is a perfectible commodity, and that flawless exchange lies just around the corner.[31] But as communication theorist John Durham Peters reminds us, "It's a mistake to think that communications will solve the problem of communication, that better wiring will eliminate the ghosts."[32] In the above example of a chronically ill patient, there is no suggestion that the original information entered into the system that would be called up by the patient's chip might already contain errors, be out of date, or be irrelevant. Instead, the assertion is that these new surveillance technologies can compel the individuals into communicating bodily truths, whether they want to or not.

This theme of perfectible communication arises again in a CNN news broadcast detailing the advantages of the implantable Verichip™.[33] Sounding more like a salesperson than a journalist, and revealing the impossibility of separating state institutions from profit motives, the CNN anchor begins by saying, "Imagine an emergency room where a doctor could scan a critically ill patient in an instant and know his or her entire medical history. It's an idea that could save lives, it also could get under your skin. Welcome to the future."[34] Following a brief flash of the text: "Denise's wish: Peace of mind," the camera then fades quickly to a woman who suffers from seizures following a serious car accident. Denise, the protagonist of this news spot, fears having a seizure far from home and ending up unconscious at a foreign hospital—unable to communicate her medical condition to the staff on duty. As a result, she is too afraid to leave the country. In discussing Verichip's implantable RFID tag, Denise suggests that for her, it might be the lifeline that would allow her to travel. No mention is made of the fact that less than 200 hospitals in the United States carry this technology,[35] making it unlikely that a hospital abroad would have the required scanner. Moreover, supposing English-language dominance, there is no discussion of the possibility that, even if a hospital were able to pull up Denise's medical records, they might not be able to read them.

Of course, in suggesting that they can compel bodies to communicate, we can see the ways that the RFID industry paves the way for imagining how these new surveillance technologies might be used to force deceitful bodies to speak. Masquerading as technologies of truth, RFID technologies are productive. They produce new forms of bodily knowledge in addition to new ways of documenting, understanding, and communicating with the human body. In particular, RFID technologies produce bodily interiors as conduits to home truths. Unlike visual surveillance technologies (including the Panopticon) that aim to flatten the body in order to render it transparent, RFID technologies produce the body in three dimensions. In opening up the body's interior, RFID technologies are used to colonize inner bodily space, making bodily interiors productive for the purposes of governance. In the following case study, we will see the ways that RFIDs are represented as able to force "othered" bodies to reveal secret information, producing bodily interiors as repositories of honesty.

IMPLANTING NEWCOMERS

Although Verichip proposed chips for medical patients as one of its first applications, it quickly became clear that the company needed to find new markets if it wished to turn a profit. By the end of 2006, only 222 medical patients had opted to have RFID chips injected into their bodies, according to documents that Verichip filed with the Securities and Exchange Commission (SEC).[36] This number is modest, and the revenue generated by implantable radio tags was far below the company's predictions.[37] Verichip set the price "for the first 3.1 million shares released in the IPO...at $6.50–8.50 per share," but they fell to $5.67–$6.99 per share within the first three months.[38] Given Verichip's financial situation, it did not take long for the company to propose expanding its technology to other programs.

Enter Verichip into the immigration debate. In 2004, when the FDA approved subcutaneous RFIDs, immigration dominated national news. The 1990s bore witness to hundreds of bills proposing regressive immigration reform, a trend that only intensified following the events of 9/11. In 2007 alone, 1,500 immigration-related bills were proposed at all levels of government.[39] Although legislation granting greater rights to immigrants and refugees was proposed, in general, the only successful legislation was that which imposed increasingly punitive sanctions against undocumented workers.[40] Many new draconian measures have been instituted in the last ten years, from allowing local law enforcement officials to enforce federal immigration law[41] to new regulations depriving immigrants of basic social services such as welfare and health care.[42] Numerous tools aimed at checking immigration status were developed as part of the crackdown on undocumented workers. The Department of Homeland Security attempted to institute what it called the E-Verify system, enabling employers to check the residence status of potential employees. With similar intent, the Real ID Act was signed into law in 2005.[43] Calling for national standards for American identification documents, the Real ID Act turns driver's licenses into de facto national ID cards that may be compared with both federal and state databases, enabling an easy check into a person's immigration status among other worrying privacy violations. Alert to the possible business opportunities, Scott Silverman, the CEO of Applied Digital, wasted neither time nor money in suggesting that implantable RFID tags might be the perfect solution to the "problem" of immigration. A year after its patent was approved, Verichip hired the firm Oldaker, Biden[44] & Belair to lobby policy makers in Washington, D.C. about their product to the tune of $120,000 a year between 2005 and 2007.[45] In 2006, Silverman embarked on a series of interviews in which he suggested that chipping all immigrants and refugees to the United States might be the answer to contemporary immigration concerns. As we see in this interview with Silverman from Fox News,[46] an implantable radio tag is marketed as a more secure, tamper-proof "passport," allowing for increased security:

> *TIKI BARBER, coanchor: All right now, could implanting a microchip into guest workers coming into the US solve our illegal immigration problem?*
> *BRIAN KILMEADE, coanchor: Here to tell us right now why this is a viable solution that might be used very shortly, Scott Silverman, CEO and Chairman of Applied Digital [Verichip's parent company]. Scott, where is this being used right now?*
> *Mr. SCOTT SILVERMAN (Chairman & CEO, Applied Digital): Well, this chip today is being used for medical applications, to identify high-risk medical patients and their medical records in an emergency and clinical situation. The chip itself was approved by the FDA several years ago as a class-two medical device, specifically for that application. But obviously, it can be applicable for the immigration issues we face today as well.*
> *KIRAN CHETRY, coanchor: But it is an interesting phenomenon. I don't know how comfortable even if you asked me or Tiki or Brian if we would be willing to do it. It just seems, like—it seems scary.*
> *KILMEADE: If I wanted to come to the United States, chip me to death!*
> *BARBER: But it really is no different than having a passport and having a way to identify yourself. This just is a way that you won't lose it.*
> *Mr. SILVERMAN: Yeah. It's a benefit to the person that's in the guest worker program, because if you leave your card at home or you leave it at your work, you're not going to be able to go back and forth across the border.*

KILMEADE: It's like permanently putting a string on your finger to remind you of something.
Mr. *SILVERMAN: Correct. That's correct.*

Although the technology here is promoted by Silverman as neutral, hardly different than a simple mnemonic device such as a string on your finger, in fact, we see how subcutaneous radio tags could reinforce existing forms of systemic discrimination. In recommending that RFID tags be implanted into immigrants and refugees to the United States, these new identification technologies are being suggested for use in one of the nation's most vulnerable communities. Although Silverman says that any use would be strictly voluntary, it is easy to imagine ways in which consent to these technologies might be compromised. Represented as able to render bodies that wish to remain silent with the power of speech, far from adding "mechanical objectivity" to the science of identification, instead radio tags are used to cement existing inequalities, including the intersection of classism and racism through compelled communication. Here, we have the outcome of assertions that these new surveillance technologies can compel the individuals into communicating bodily truths, whether they want to or not, as implantable RFIDs are represented as able to productively fill bodily silences with the speech of othered bodies.

Ultimately, Verichip™ was not successful at promoting its implants for use in immigrants and guest workers. Those wishing to implement new surveillance technologies are always in conversation/tension with those resisting their use. Consumer advocacy organizations, privacy advocates, fundamentalist Christians, as well as antipoverty and progressive immigration activists all resisted the development and implementation of RFID chips. Particularly successful protests were staged by the consumer advocacy organization CASPIAN. CASPIAN resists all RFID tags—which they have usefully termed "spychips"—including both subcutaneous chips as well as those used to chip inventory.[47] In addition to using tactics such as protests and boycotts, CASPIAN also drew on scientific research calling into question the ability of these technologies to work reliably. CASPIAN highlighted that early tests on mice demonstrated the detection of elevated levels of cancer after the chips were injected. In addition, though not specifically related to the suggestion that immigrants and guest workers be implanted with chips, the landmark 2006 immigration protests in the United States undoubtedly had an effect on the imagined possibilities for Verichip's subcutaneous products.

Due to a combination of factors, including evidence that the technologies could be easily hacked,[48] CASPIAN's successful consumer advocacy protests,[49] as well as research indicating that they might pose health risks,[50]

Verichip's implantable radio tags have not sold well. As Verichip stated publicly in 2007, "To date, we have only generated approximately $0.1 million in revenue ($100,000) from sales of the microchip inserter kits, significantly less than we had projected at the beginning of 2006. We may never achieve market acceptance or more than nominal or modest sales of this system," the company stated.[51] Through the study of the attempted application of RFID tags to newcomers to the United States, we see how a conversation between powerful corporate and state interests attempted to silence certain kinds of discussions about substantive equality while simultaneously filling vulnerable bodies with a new kind of speech—one that locates bodies temporally and spatially in troubling ways. However, we also see that resistance to these chips reveals the ways that RFIDs remain caught in a field of struggle and contestation between those in power and vulnerable groups. Moreover, the success of the opposition to RFID chips as a result of these rich and varied forms of resistance to them tells us as much about the productive possibilities for organized struggles formed from alliances across difference as it does about the possibilities for opposing new surveillance technologies.

And yet, Verichip's implantable tags are far from dead. Following Hurricane Katrina, in an attempt to speed the identification of corpses by cross-checking chipped bodies with databases of who had died in a particular geographic area, approximately 300 bodies were implanted with Verichip™ technology[52]—compelling these bodies to speak their identities from beyond the grave. In addition, VeriKid programs were started in both Mexico and Brazil. Verichip markets implanted chips for children as "good" protection against kidnapping; Mexico's VeriKid program is backed by the country's "National Foundation of Investigations of Robbed and Missing Children."[53] Most recently, proposed legislation in Indonesia's Papua region suggested that every HIV positive person be implanted with a chip.[54] Although Indonesia's legislation was tabled after an international outcry, it is clear that we have not seen the last of these chips, as they continue to be proposed for the most vulnerable segments of the population in ways related to systemic forms of nationalism, ageism, racism, classism, and ableism.

CONCLUSION

The body under surveillance is one that state and corporate institutions and industry aim to make perfectly visible, shining a light into all of its hidden corners. And yet, my analysis of subcutaneous radio tags reveals that new surveillance technologies do not stop with making the body transparent. Circumventing the body's surface, these implanted tags also are used to make the body audible. Subdermal RFID tags link bodies into communication networks in ways that allow institutionalized forms of power to compel formerly silent bodies to speak.

In historicizing the development of RFIDs, from the beginnings of using radio tags to chip people in Guantanamo Bay to the use of implanted tags in Alzheimer's and other chronically ill patients to the suggested use of implanted tags in immigrants and refugees, an examination of their origins clearly demonstrates the ways that they reinforce contemporary forms of inequality. Radio tags, previously used primarily in consumer applications such as inventory control, are now being used to suggest the possible construction of individuals as human inventory as these technologies are used to make bodies newly locatable in time and space.

Corporate profiteering is central to the expansion of RFIDs. That is, the attempted expansion of radio tags is linked to industry desires for new markets. Clearly, the easiest way to increase profits is to find new populations to tag, and Verichip™ has reliably suggested the use of its chips on the most vulnerable segments of the population, who are additionally those least capable of refusal. The attempted application of RFIDs to immigration also reveals that powerful corporate interests wishing to find as broad a market as possible for their technologies are always in conversation with those resisting their use. In large part as a result of both separate and coordinated efforts aimed at halting RFID expansion, Verichip has been largely unsuccessful at applying its technologies to newcomers to the United States.

The cultural contexts driving the development of RFIDs, the use of communications technologies as policing strategies directed at vulnerable and marginalized groups, and the interrelationship of industry demands and communications policy are all central to understanding the growth of implanted radio tags. We must continue to question the narrative of technological progress articulated to promote the adoption of subcutaneous RFID chips and to demonstrate the ways in which they endanger democracy, socially just forms of subjectivity, and substantive forms of equality.

NOTES

The author would like to thank Tara Rodgers and Corinne Mason for their terrific research assistance, as well as Darin Barney (postdoc supervisor extraordinaire), Paisley Currah, and Monica Casper for their excellent comments on an earlier version of this chapter. I would also like to thank Sarah Berry, Ummni Khan, Suzanne Bouclin, Kathryn Trevenen, Michael Orsini, Simone Browne, Carrie Rentschler, Donna Haraway, Celiany Rivera Velazquez, Robert Smith, Lisa Nakamura, Paula Treichler, C. L. Cole, Kent Ono, Helen Kang, Aisha Durham, Amy Hasinoff, Carolyn Randolph, and Himika Bhattacharya for technical discussions that helped to inform my thinking in this chapter.

1. Mikey Sklar, "Future Shock: The Daily Show," ed. Jon Stewart (2006), http://vimeo.com/2294067.

2. Simone Browne, "Getting Carded: Border Control and the Politics of Canada's Permanent Resident Card," in *The New Media of Surveillance*, eds. Shoshana Magnet and Kelly Gates (New York: Routledge, 2009).

3. Shoshana Magnet, "Encoding the Body: Critically Assessing the Collection and Uses of Biometric Information" (University of Illinois at Urbana-Champaign, 2008); Paula A Treichler, Lisa Cartwright, and Constance Penley, *The Visible Woman: Imaging Technologies, Gender, and Science* (New York: New York University Press, 1998); Lisa Cartwright, *Screening the Body: Tracing Medicine's Visual Culture* (Minneapolis: University of Minnesota Press, 1995).

4. Lorraine Daston and Peter Galison, "The Image of Objectivity," *Representations* 40 (1992).

5. Rob Stein, "Bar Code Implant Calls up Medical Data," SF Gate, October 14, 2004, http://www.sfgate.com/cgi-bin/article/article?f=/c/a/2004/10/14/MNGQA99FDM1.DTL.

6. Katherine Albrecht and Liz McIntyre, *Spychips: How Major Corporations and Government Plan to Track Your Every Move with Rfid* (Nashville, TN: Nelson Current, 2005).

7. Kim Zetter, "School Rfid Plan Gets an F," *Wired*, October 2, 2005, http://www.wired.com/politics/security/news/2005/02/66554.

8. Jo Best, "Japan School Kids to Be Tagged with Rfid Chips," ZDNet, July 13, 2004, http://news.zdnet.com/2100-3513_22-137122.html.

9. Claire Swedberg, "L.A. County Jail to Track Inmates," *RFID Journal*, May 16, 2005, http://www.rfidjournal.com/article/articleview/1601/1/1/.

10. Jim MacKay, "Prisons Use RFID Systems to Track Inmates," Government Technology 2008, http://www.govtech.com/gt/312938?topic=117680

11. "Verimed Patient Identification System Expands to Emergency Medical Responders," *Business Wire*, November 14, 2007, http://www.allbusiness.com/health-care/health-care-facilities-nursing/5311797-1.html

12. Verichip is a subsidiary of Applied Digital. Applied Digital sells a broad range of services aimed at computerized automation and control.

13. Verichip, n.d.

14. Ibid.

15. Lu Yan, *The Internet of Things: From RFID to the Next-Generation Pervasive Networked Systems* (New York: Auerbach, 2008).

16. Beth Rosenberg, *RFID: Applications, Security, and Privacy* (Upper Saddle River, NJ: Addison-Wesley, 2006).

17. Albrecht and McIntyre, *Spychips*.

18. Simon Cole. *Suspect Identities: A History of Fingerprinting and Criminal Identification* (Cambridge, MA: Harvard University Press, 2001).

19. Yan, *The Internet of Things*.

20. Rachel Hall, "Of Ziploc Bags and Black Holes: The Aesthetics of Transparency in the War on Terror," *The Communication Review* 10, no. 4 (2008).

21. Treichler, Cartwright, and Penley, *The Visible Woman.*

22. Also relevant to theorizing the body under scrutiny is research within the growing field of surveillance studies. David Lyon's work on thinking about surveillance as a process of social sorting is particularly useful to analyzing the relationship between surveillance and bodily identities (2003), as is research that addresses the relationship between surveillance and social inequality more broadly. Richard Ericson and Kevin Haggerty, *Policing the Risk Society* (Toronto: University of Toronto Press, 1997); Torin Monahan, *Surveillance in the Time of Insecurity* (New Brunswick, NJ: Rutgers University Press, 2010); Kelly Gates, *Our Biometric Future: Facial recognition technology and the culture of surveillance* (New York: NYU Press, 2011); Greg Elmer. *Profiling Machines: Mapping the personal information economy* (Cambridge: MIT Press, 2004); and Lisa Nakamura, "Interfaces of Identity: Oriental Traitors and Telematic Profiling in *24*," *Camera Obscura* 24, no. 1 (2009).

23. "25 Top Influencers in the Rfid Industry," RFID Gazette, February 20, 2007, http://www.rfidgazette.org/2007/02/25_top_influenc.html.

24. Yan, *The Internet of Things.*

25. Albrecht and McIntyre, *Spychips.*

26. VeriChip, n.d.

27. Ibid.

28. Although it is beyond the scope of this chapter, further work might investigate how RFID tags are used to reinforce understandings of livestock as no more than sentient inventory.

29. Paula Treichler uses the term "sightbyte" to describe those representations that have a dramatic visual impact in ways that work against more complex or nuanced readings of a particular event, such as the collapse of the twin towers on 9/11 (personal communication, 2008).

30. Paula A. Treichler, *How to Have Theory in an Epidemic: Cultural Chronicles of Aids* (Durham, NC: Duke University Press, 1999).

31. As Paisley Currah usefully notes, these dreams of transparent communication date back to Enlightenment desires for perfectible language. Of course, any form of communication still requires an educated and equipped reader/listener.

32. John Durham Peters, *Speaking into the Air: A History of the Idea of Communication* (Chicago: University of Chicago Press, 1999).

33. CNN, "Welcome to the Future," (United States 2006), http://www.verichipcorp.com/content/media/audio_video.

34. Ibid.

35. Tricia Cassidy, "Chipping Away at Health Information Technology," Verichip: News, September 1, 2006.

36. Michael Kanellos, "Patients, Doctors Staying Away from Implantable Rfid Chips," CNet News, February 12, 2007, http://news.cnet.com/Patients,-doctors-staying-away-from-implantable-RFID-chips/2100-11746_3-6158701.html.

37. Ibid.

38. Ibid.

39. Alan Greenblatt, "Immigration Debate," *CQ Researcher* (2008).
40. Ibid; Kent A. Ono and John M. Sloop, *Shifting Borders: Rhetoric, Immigration, and California's Proposition 187* (Philadelphia, PA: Temple University Press, 2002).
41. Greenblatt, "Immigration Debate," #35.
42. Ono and Sloop, *Shifting Borders.*
43. Greenblatt, "Immigration Debate."
44. R. Hunter Biden, the cofounder of this firm, is the son of Joe Biden.
45. OpenSecrets.org, "Annual Lobbying by Verichip Corp," http://www.opensecrets.org/lobby/clientsum.php?year=2008&lname=VeriChip+Corp&id=.
46. Tiki Barber and Brian Kilmeade. Fox News, May 16, 2006, http://www.spychips.com/press-releases/silverman-foxnews.html.
47. Ibid.
48. Jonathan Westhues, "Demo: Cloning a Verichip," 2006, http://cq.cx/verichip.pl.
49. Albrecht and McIntyre, *Spychips*
50. Erika Morphy, "Possible Rfid-Cancer Link Rattles Market," *Tech News World*, November 9, 2007.
51. Kanellos, "Patients, Doctors Staying Away."
52. Alorie Gilbert, "Rf-Iding the Dead," *CNet News*, January 12, 2006, http://news.cnet.com/RF-IDing-the-dead/2008-1006_3-6017623.html?tag=nw.6.
53. Julia Scheeres, "Tracking Junior with a Microchip," *Wired News*, October 10, 2003, http://www.wired.com/science/discoveries/news/2003/10/60771.
54. "Indonesia's Papua Plans to Tag Aids Sufferers." Reuters, November 24, 2008, http://www.reuters.com/article/healthNews/idUSTRE4AN3U620081124.

CHAPTER 10

VIRTUAL BODY MODIFICATION: EMBODIMENT, IDENTITY, AND NONCONFORMING AVATARS

MARY KOSUT

We are encased in lovely and unlimited avatars.

—Second Life forum

Media are extensions of the senses.

—Marshall McLuhan

INTRODUCTION

Computer-mediated interactions permeate and influence the daily lives of an increasing number of people, as evidenced in the meteoric rise of social networking sites such as MySpace and FaceBook. Online experiences have been transformed thanks to Internet-based technologies that are increasingly interactive and user-friendly. For example, blogging and sites such as Flickr, Twitter, and Wikipedia are part of a new wave of collective, interactive or user-generated media. Such technologies, often lumped into the category of "new media," have changed the way we communicate and gather information, yet they also shed light on the complex relationship between mediated or virtual interactions and those that are in the flesh.

Recently, complicated connections between real and virtual interaction were brought to the world's attention vis-à-vis a popular news story, "Virtual

Affair Ends in Real-Life Divorce,"[1] which chronicled the romantic lives of a
British couple who spent much of their time socializing in the online virtual
world Second Life. While married both in real life and in Second Life, the
husband was caught virtually cheating on his wife with a "pixel prostitute."[2]
While salacious and tabloidesque, the overlap between the real and virtual
lives of this couple is nonetheless instructive; it points to how technology
intersects and may change people's intimate lives.

Powerful emotional and social connections are not specific to new media
forms. However, contemporary virtual social environments (such as Second
Life) and role-playing games (such as World of Warcraft) facilitate a particular intimacy that is often not only emotional, but also physical in nature.
People interact with such sites not only with their bodies, but *through* their
bodies as well. That is, they are so closely and emotionally attached to their
avatars (self-representations) that online interactions do not simply transpire
"out there" in an abstract virtual space. Virtual social activities can be characterized as visual, imaginary, or fantasy-based, but they are simultaneously
lived, in a sense, *brought to life*, from within corporeal bodies that feel, desire,
crave, and react to the stimuli and situations they encounter.

Every new medium, whether in the form of an iPod, mobile phone, or
computer, engages the body in distinctive ways. Often such technologies
become part of our everyday embodied routines, suggesting a type of mundane "cyborgification."[3] Technology so infuses our lives that its absence
is considered unthinkable (i.e., we can't live without it). As media scholar
Frank Biocca noted over a decade ago:

> The process of progressive embodiment is occurring at a time when there is
> increasing social integration of the interface. Social integration means that
> the interface is being integrated into everyday activity at work, home and
> on the street... [it] enters the social sphere via easier coupling with the body
> through miniaturization, portability and wearability.[4]

Although Biocca was referring to virtual reality (VR) technologies such as
data gloves and helmets, the union of user to "interface" is clearly evident.
New media technologies are fused with our habits, routines, and the very
contours and surfaces of our bodies.

Computer-mediated interactions in particular provide a new space to
revisit key questions regarding the nature of embodiment and the connections between bodies and media. The former points to how we experience
life from within our own bodies in a phenomenological sense, and the latter
intersects media, representation, and identity (and the inherent extension
of this triad to the body). In this chapter, I explore the messy, interactive
relationship between "real" flesh and blood bodies and computer-generated

visual representations of people, or avatars. How do these figures reflect and connect to the real-life body of the user? In what ways do avatars both conform to and subvert bodily norms?

In considering these questions I draw from my own experiences navigating Second Life, as well as blogs within and about Second Life. I focus on Second Life specifically because avatars in this cyber environment represent people in real time. It is a distinctly interactive social site that illuminates the complexities between dichotomous terms such as real and virtual, digital and actual, and so forth.

"IT DOESN'T REPLACE YOUR REAL LIFE. IT MAKES IT BETTER"[5]

Second Life (referred to as SL for the remainder of the chapter), launched in 2003, is a virtual world owned by Linden Labs. Today, there are approximately 1.5 million registered users. SL has many cultures and communities and its own economy (which uses Linden dollars) as well as a distinct time zone. A "basic membership" to SL is free to any user and allows participation in all activities except for land ownership. Anybody who has the ability to use a computer, can see a keyboard and screen, and physically press keys or control a mouse can use SL. This obviously excludes people with certain physical disabilities who are unable to afford technological modifications that can facilitate SL involvement. Using SL does require an investment of time to understand how to create and modify avatars, move them through the endless spaces and places available, and interact with other avatars via text or a range of physical gestures such as winking, laughing, blowing kisses, and so forth.

Those who live part of their virtual lives in SL may also participate in other virtual environments such as World of Warcraft (which is combat-oriented) or Entropia Universe (in which participants seek to colonize the planet "Calypso"). In comparison to SL, both World of Warcraft and Entropia Universe differ in that they revolve around game interaction. Game worlds, like classic "video games," are "objective driven systems in which a player tries to achieve a goal under a proscribed set of rules."[6] Participants strive to learn how the game works and are in a sense bound within the parameters of the game world itself. Comparatively, SL is a distinctly social world in which users create the rules, roles are less fixed, and intensive social interaction takes place across a plethora of user-built and imagined spaces, communities, and activities. However, systematic differences aside, people who avidly use virtual worlds and games are often highly invested in the norms and rules that frame interaction (regardless if they are fixed or emergent). For example, "Doug," a forty-something enthusiastic daily user of Entropia

Universe told me, "We know who belongs in the game and who is there for the wrong reasons. You can tell when someone doesn't fit in with the society." In this sense, virtual reality mirrors the firmness and structure found in the everyday world.[7] There is maintenance of norms, a semidefined social order, and disciplinary regulators that govern emergent social interactions.

Some people choose SL avatars based on fictional characters, Hollywood celebrities, or even nonhuman representations like The Furries, a popular community of users who virtually embody furry animal avatars. However, Kristine Nowak and Christine Rauh suggest that when it comes to the variable of gender, people prefer avatars that are like themselves.[8] In addition, in the case of images of static avatars, representations that were most feminine and anthropomorphic were believed to be both more credible and attractive by the participants.[9] Preferences aside, I focus exclusively on avatars that are idiosyncratically but distinctly human in appearance. Within this classification a vast majority of SL avatars conform to hegemonic norms of masculinity and femininity, often in the form of extreme representations of both. Female avatar bodies are overwhelmingly elongated and thin, with tiny waists, gigantic breasts, long hair, and full lips. Male avatars, not surprisingly, tend to have broad shoulders, strong jaw lines, and bulging, well-defined muscles.

Besides constructing gender, it is possible to racialize one's avatar by adjusting skin tone, facial features, and hair. Skin color can be adjusted via SL controls, but "skins" are also for sale offering the possibility of sundry shades and textures. Notwithstanding, the vast majority of avatars have lighter skin tones. While conducting research online I do not recall seeing a humanesque avatar that had dark skin or could be identified as racially nonwhite. Just as binary gender norms prevail within on and offline cultural milieus, so too does the privileging and ubiquity of lighter skin and features associated with or defined as "white."

However, my primary focus is not on avatars that conform to standard body hegemonies, whether in the form of gender norms or racial stereotypes per se, but those that are *nonconforming* with respect to societal standards that define what is and is not normal or beautiful. Nonconforming avatars are those that are purposefully constructed as different by their users: they may be heavier, smaller, shorter, or exhibit evidence of having a physical disability through the use of a virtual wheelchair, cane, seeing-eye dog, or prosthesis. Of course, there is no way to verify whether the person who constructs his or her avatar looks similar to or different from his or her avatar in real life, or is similarly abled or disabled. SL allows room for play, readjustment, irony, and subversion. There is a shifting back and forth between fantasy and reality within a simultaneously virtual and material media interface. This is what media scholars refer to as "the bleed."

VIRTUAL ETHICS AND METHODS

Participating in data collection as an avatar, as well as analyzing the content of blogs, raises a number of ethical issues related to virtual media ethnography, and the study of new media technologies and their users more broadly. As social worlds, blogs, role-playing games like MUDS (multiuser domains, text based) and MOOS (multiuser domains, object oriented), and social networking sites have flourished, scholars have grappled with how to study and collect data within cyberspaces. Arguably, media researchers from across disciplines are still catching up with a diverse evolving media terrain. In 2002, AoIR, the Association of Internet Researchers, formed an online ethics working group that drew up a comprehensive and detailed guideline for those working in/with a variety of media platforms. While the AoIR ethics report covers a range of settings for study (e-mail, MUDs, chatrooms, etc.) and is sensitive to the dynamics of the cultural backgrounds of the researchers themselves, it does not provide a definitive set of rules.

As this chapter is based primarily on the textual content of blogs about SL that are public and permanently archived, and there is no password required to access the information or policy that prohibits such research, it is public data (much in the same way a *New York Times* article is archived and accessible). Yet, these texts were authored by private citizens and are a reflection of their personal experiences within SL. Even though the content is technically public, it is about, and for, a specific community of others. With this in mind, I became what some call "a lurker"; I am not an avid SL user, and I did not post messages in SL forums. I simply read and then analyzed the data.

In addition, I also "lurked" as an avatar in SL, engaging in virtual participation, observing others and the various activities that take place within the site. However, because I did not reveal my identity to others in SL and visited in a purely exploratory fashion, I do not refer to specific conversations with SL avatars within this chapter, only to their online posts and my *own* experiences with/as an avatar.

Even though consent was not required as per AoIR ethics guidelines, I agree with sociologist and cyber scholar Dennis Waskul that "responsible or ethical research is not a matter of codes, policy or procedure. Rather, responsible and ethical research centers on a commitment to protect the participants of one's study."[10] With this in mind, in order to protect subjects' identities, I changed the screen names of those who posted comments as well as the name of my own avatar, just as it is common to change an informant's name within a traditional ethnographic study. Because the boundaries between public and private are so murky in online research, it may be the case that "even if a certain internet medium admittedly *is* public, it doesn't *feel* public to its users."[11] In addition, many users are aware that others can either read their commentary or watch their virtual interactions, but they likely do not consider that a researcher may be documenting it as data to be analyzed. SL users engage in intimate activities and relationships and they are forthcoming in their discussions of such intimacies in forums and blogs, whether the topic is virtual sex or the close, personal embodied connections some establish with their avatars. Personal experiences such as these can potentially cause emotional harm for users, in the form of embarrassment or shame, when they are read in different contexts by those who are not privy to SL norms.

In addition to ethical issues, another factor that cyber researchers have grappled with is the distinction made between face-to-face data collection and that which is technologically mediated. As Christine Hine observes, "Methodological solutions do not automatically transfer from offline to online settings."[12] However, a binary online/offline distinction (the latter as somehow more real or valid than the former) has been repudiated and challenged by a number of scholars.[13] A rich and rapidly developing body of work from cyber researchers within the social sciences and media and communication studies highlights the "realness" of online data and the intersections between online and offline social interactions.[14]

MODIFYING THE VIRTUAL BODY/SELF

Body modification practices and regimes have become a taken-for-granted aspect of everyday life in the twenty-first century.[15] The rise in tattooing, cosmetic surgery, alternative health treatments, and exercise and diet programs signal a historical moment in which bodily maintenance is normative. As a primary source of identity, many people with discretionary time and income actively work on their bodies; adjusting, improving, rewriting, honing, adorning, and so forth. Bodily appearance can be read as a form of physical capital, which can be converted into both economic and cultural capital.[16] Not surprisingly, because the SL avatar represents the virtual body of the user, users often spend a considerable amount of time (and money) working on their avatars' bodies, adjusting the height, shape, clothing, skin color, hair, and so on. This is evident in the complexity and detail of some avatars that/who wear elaborate clothing and accessories or are radically modified.

As discussed previously, in SL life just as in real life, appearance is of great consequence. However, SL offers more room for gender play and nonconformity. Some users engage in virtual drag, as men can virtually embody female avatars and vice versa. And of course, avatars can be constructed as purposefully androgynous, or simultaneously represent conflicting gendered physical cues and symbols. The connection between the avatar's image and fixed social status variables, such as gender, race, and sexuality, do not always directly correspond within a creative and fluid environment such as SL. Some have asserted that cyberspace offers greater opportunities to exist in a nuanced gendered continuum where masculine and feminine can collide, overlap, or be resisted altogether.[17]

Yet, visual binary gender systems often seem to be reinforced in many video games, 3D chat rooms, and consumer-driven interfaces. A cursory glance at video games in particular reveals an almost extreme exaggeration of gendered physical characteristics, as in the well-known example of Lara Croft Tomb Raider. Despite the endless opportunity for innovation, creativity, and transgression, some argue that the borders between masculine and feminine are not being rewritten, but instead uncritically upheld and even inflated in cyberspace.[18] Notwithstanding, these gendered exaggerations may be purposefully employed with a sense of irony and play.

The term "presence" refers to the physical sensations experienced while engaged in virtual reality or with media generally. According to Frank Biocca, "Users experiencing presence report having a compelling sense of being in a mediated space other than where their physical body is located."[19] Evidence suggests that some users both feel and think *with* and *through* their avatars. Although some scholars have been dismissive of the idea of physical online embodiment or bodily presence, there are those who assert that presence is experienced within the body.[20] The user is present cognitively as a thinking reflexive self and is engaged within a particular cultural context. The dimensions of self, and society as it were, extend to/through the body. This mediated self/body/society triad rejects the Cartesian binary of body as distinct from self. The connection between avatar and body/self is described by SL-user Medina Rundgran who writes, "Just like in RL I think my breasts are a bit on the small side. Copying this in a truthful way makes me feel very connected with my avatar in intimate situations." Medina's avatar mirrors her own physical body, and thus allows her to "feel" virtual sexual interactions in a tangible physical and emotional way.

When it comes to the visual appearance of avatars and their relationship to the fleshy bodies of their users, some report a direct connection between the SL and RL body (posts often use "RL" in place of real life). This is illustrated in commentary by "Blue Lilly" who writes, "She is me, I am her, etc. I am very thin in RL due to the sport that I am competitive in and practice everyday. I am also relatively tall for a woman at almost 5'7." Like "Blue Lilly," "Mimi Syncata" also described her avatar as being directly connected to her RL body/self: "After about 10 shapes, I found one with facial features I liked that also had a nice curvy shape with ample booty and thought it suited 'me' best." Finally, "Winifred" makes an even more direct link between herself and her avatar, "Rosie," who she personifies:

> Rosie and the RL me have come together in many ways since I started SL 10 months ago. I've started looking and dressing like her (I actually bought a pair of black pants my husband calls the "avatar pants"), and she more like me, especially in coloring....I find it interesting though that in a world where you can literally be whatever you wish, most people myself included are pretty similar in form.

For certain individuals, the lines between the avatar and physical body/self are blurred. Both literally and conceptually, a triadic avatar/body/self is constructed.

Science and technology studies scholar Steve Woolgar's work on the virtual is germane with regard to avatar construction and the body/self of the user.[21] In his "five rules of virtuality," he emphasizes a new way of thinking about the social dimensions of new technologies, proposing that the more virtual, the more real.[22] What this suggests is that virtual interactions are indeed "real" activities. The body and the interface are *one and the same.*

CONSTRUCTING THE AVATAR: PRUDENCE AND I

I became a member of the SL community on May 17, 2008, when my avatar "Prudence Pentanque," emerged on Visitor's Island, which is the first area you enter after logging on. Upon arrival, most avatars' appearances are very basic, confined to stock shapes and accessories (referred to as "prims"). However, as one gains more knowledge of the tools and options available, avatar editing possibilities are limitless. Virtual body modification is highly creative, as the form of the avatar (skin color, nose, hair, eyes, hands, etc.) or avatar's body, and the outfit it wears, are contingent on the user's imagination (and ability to navigate the controls). For example, you can choose clothing style, as well as the texture of a particular fabric, such as linen, angora, or denim. The appearance of one's avatar, much like the presentation of self in everyday life, is important in defining virtual interactions between and among others.[23] According to *Second Life: The Official Guide,* "Your avatar choice says a lot about who you are; to the people you encounter in the SL worlds, your avatar *is* who you are. It's true too—your avatar choices reflect your personality and mentality. It's good to keep that in mind.[24]" Interestingly, this quote is underneath and refers to "Figure 1.3" which is an image of a bald, fat, white male avatar in jeans and a t-shirt with the caption "I wonder if that's really me."[25] Novice users are advised to consider the attractiveness of their virtual

Figure 10.1   Prudence Pentanque.

body and implicitly warned of the ramifications of constructing an avatar that is nonconforming. There is, apparently, a right way to virtually modify your avatar's body.

After a few months and countless hours invested in tweaking proportions, clothing, facial features, and hairstyle I began to refer to my avatar as "Prudence" or the pronoun "her." Interestingly, the more time I spent on her appearance, the more I wanted to show her off. Just as people often view the body as a project to be worked on, honed, and modified as a site of identity, Prudence became a reflexive virtual body project.[26] The more time I spent with/as Prudence, I began to better understand how the boundaries between the real and virtual body/self are porous. If users invest time in honing, adorning, and tweaking their avatars, it makes sense that some may become deeply attached to these representations. However, the avatar, unlike the carnal body, is infinitely modifiable and can be radically updated daily, if not hourly.

After becoming aware of the normativity and simplicity of Prudence's physical appearance, I enlisted a friend who is a SL aficionado to help me buy Prudence some unusual body parts. Outfits and bodily attachments, such as tattoos, jewelry, "skins," S & M gear, fingernails, and sunglasses can be purchased in SL stores with Linden Dollars (this is a very partial list). I did not want her to conform to actual world embodied norms and standards because virtual reality promised the seductive possibility that I/she could be *anything*. I wanted her to be distinctive and unconventional, by both SL and RL standards. Prudence is conforming in that she is tall and thin, but she has a tattooed chest and wooden peg legs from the knee down. In this way, she is a semiconforming avatar.

After envisioning and creating Prudence's form, I had to learn how to successfully move or transport her through various locations and situations, such as an art gallery, a strip club, and a library lecture hall. Although initially challenging, once you learn the basics of avatar locomotion, you can run, walk, or fly with the touch of a keystroke. This is very interesting in that the avatar must be programmed by the user's body to behave in an authentically embodied way, even if only in terms of mundane functions like waving or nodding. Mastery is important—new users (myself included) struggle to get their avatars to respond and move accordingly. I realized I was not alone as evidenced in a lecture I observed sponsored by the "gimp girl" community, a SL group that consists of people with disabilities, both actual and virtual, as well as academics that engage in disability studies. Attending the first lecture proved to be incredibly difficult for Prudence, as I was unable to do basic things such as make her sit, turn around, and walk in a purposeful direction without bumping into other avatars. While some avatars were sitting and standing with ease, one with a seeing-eye tiger and another in a wheelchair, I/she was awkward and clumsy (although nobody seemed to notice). But it was clear that others were having difficulty maneuvering and communicating during the lecture.

This experience highlights the issue of the possibility of control over the avatar's body, as well as the relationship between mind (specifically will and intent) and the execution of specific actions through the virtual body. Controlling the avatar's body is contingent not only on intellectually comprehending which keys to press to engage in a desired activity, but importantly, a physical familiarity with the keyboard. Thus, the relationship between the avatar and user's body is developed over time and is of a distinctly phenomenological nature as posited by French philosopher Maurice Merleau-Ponty. For Merleau-Ponty, perception is only possible through a carnal body that is embedded and acting within a sensual world.[27] As the fleshy incarnate body develops an innate knowledge of locomotion, the avatar's body in a sense *comes to life* in a more direct and skillful way. As one user, "Dandelous Kisco" puts it, "If one (has) to think if I press the key to make the avatar do that (then) communication (and identification) will be much weaker than when action and reaction are intuitively connected." The more we become familiar with technology, and the more effortlessly we use it, then we begin to store and draw on embodied knowledge. The knowledge the body holds is unspeakable in that it cannot be communicated sociolinguistically, but it exists. For example, our fingers may learn to quickly and efficiently send a text message over time, adapting to the necessities of the media (cell phone). The performance with and through the avatar—although technically a representation—also manifests within the body of some users, who may be emotionally *and* physically attached to their avatars. In some cases, the avatar is me—body and soul.[28]

NONCONFORMING AVATARS

Importantly, and key to my argument, is that not all SL users create avatars that resemble or mirror their physical selves. Posts indicate that some take advantage of the infinite potential of virtual body modification and construct avatars that are improved versions of themselves as per standard norms of masculine/feminine beauty. For example, "Candycane" said "I'm not happy being larger sized in RL. So I'd rather be what I want in SL." Similarly, "Eve Ansler" posts that "part of the SL fantasy for me is that I can be slim, with no wrinkles and perfect hair." These avatars, and many others, conform to hegemonic ideals of thinness.

Notwithstanding, other SL narratives show that users reflexively consider nonconformity and how their bodies may be read both offline and online. "Zena Star" describes the complex relationship between her material body, her avatar, and how she chose to play with a representation that is unconventional by SL standards:

> In RL I am taller than average, a solid 14 and curvy. I have always been curvy, even at my skinniest and fittest. When I joined SL, I decided to see what it was like to be the exact opposite: short and slender....Its amazing how the shape defined who my av became. She is an elfish, playful, friendly thing (strange, as I am quite reserved and sarcastic in RL). People on SL have even commented that I'm too short and my boobs aren't big enough. And strangely, I became proud of being different (even though my av is still clearly ideally skinny by today's standards).

Even though "Zena" is "tall and curvy" in RL and her avatar is small and "elfish," both her virtual and material bodies do not perfectly conform. Differences aside, when "Zena's" SL body changed, its/her personality also changed to coincide with the new form. This is reminiscent of those who radically modify their body via cosmetic surgery or weight loss and report being a new person; more outgoing, confident, or interested in new activities.

Just as in RL, women are to a certain extent influenced by unreasonable standards of beauty and are similarly reflexive about how their SL bodies appear. Numerous posts spoke critically and despairingly of the ubiquitous "Barbie" avatar and even of "avatar anorexia" as the body/height proportions of female avatars are thinner than RL model standards. "Cassieopia," wrote of how she realized her avatar was grossly exaggerated and how she adjusted it:

> I finally realized how skinny I looked when I found a click tool in a dress shop somewhere that told me my height and weight, 6'2"and 120 pounds. I thought I looked good because I was measuring myself against everyone else...So I widened my hips, beefed up my thighs, lowered my boobs, got shorter...I am happier with my body...I realize part of the draw of SL is fantasy...(but) I think I'd enjoy it more if it were "me" pursuing the fantasies.

"Cassieopia's" narrative is similar to "Zena's" in that she deliberately transformed her avatar to appear as nonconforming. Yet, the bleed between a "fantasy" body and the material body/self is another example of how an avatar is not "it" but "me."

Along the continuum of nonconforming avatars, disabled SL users bring into particularly sharp focus the bleed between virtual and material embodiment. The notion of ability and disability, like other dichotomous categories explored throughout this chapter, is quite thorny. Some people have obvious physical differences that are readable in the sense that their bodies provide visual cues such as a limp, speech impairment, or locomotion via a wheelchair. Yet, such individuals may reject the label "disability" for political and personal reasons. For others, the lines between ability and disability are more complicated, as bodies may break and heal, become ill and recover. As "Stone Khalo" states for some physically impaired users, "The body is no more nor less a tool than an online avatar...and is more reliable, expressive and liberating" than the real body, which breaks down, is unreliable and sometimes recalcitrant.

Although time must be invested to learn how to maneuver the avatar with ease, physically disabled users report that avatars can become enabling (once you have figured out the features), providing a sense of freedom. One aspect of freedom discussed is physical independence or the ability to move the avatar through virtual spaces with little effort. "Kino Xwret" made a comparison between his RL and SL body in this regard, "Standing is an effort of will. Walking is an effort of will. Walking 'normally' is sometimes a severe effort of will. But in SL, things that are painful and difficult in RL are a single keystroke." SL bodies offer the liberating possibility of being in control. "Peet Fhantom" links control with a sense of attachment: "If you feel much more in control of your body than you do your avatar, you will feel a stronger attachment to your body. Likewise, if you feel more in control of your avatar, you will feel more attached to it." This suggests that an avatar can be particularly meaningful to a user whose body is difficult to predict, manage, or control.

Besides the ability to move and interact with ease, and in similar ways as other avatars, SL can afford disabled users the freedom to define their visual presentations of self. Some do choose to appear similar to their RL self/body, including using a virtual wheelchair or cane, but others may inhabit a body that deliberately conforms. As "Gardinia" observes, "It is a perfect place for anyone, anywhere, to go in and actually feel accepted, instead of being the odd one out which is most likely the case in a RL situation." In some cases, the body may be nonconforming in RL, but may conform to the avatar appearance norms in SL. Whether conforming or nonconforming visually, avatars enable their users, able or disabled, to transgress the boundaries and limitations of RL norms and bodies.

CONCLUSION

The reflexive and embodied experiences of SL users provide fruitful terrain to consider the notion of a "lived body" in the context of new media.[29] What this means is that the body is not reduced to biological categories or social variables, but conceptualized as an entity that experiences and feels the world sensually and carnally. We must continue to investigate how technologies shape how we envision what a body is, and ways in which media interfaces can enhance and extend our everyday lived embodied experiences.

The continuum of nonconforming avatars, whether it is a woman who rejects the Barbie ideal, an able-bodied person who embodies a disability, or vice versa, points to how people have absorbed the norms of RL, but use SL as a platform of resistance, subversion, and play. Conceived of as a space of transgression and interplay between ideal and real world norms, SL and other virtual worlds offer a novel way to reconsider the relation-ship between media, representation, identity, and embodiment. Instead of theorizing new media within the context of a dichotomous real versus vir-tual body/self framework, the complexity of nonconforming avatars can be used to reconsider dominant discourses and frameworks. What our culture deems "normal" and "ideal" speaks to how we consider and position the "real" and the "virtual," all of which are, of course, socially and historically relative.

As disability studies scholar Lennard J. Davis observes, "In a culture with an ideal form of the body, all members of the population are below that ideal."[30] He demonstrates how it is the construction of normalcy within cultural texts and discourses, not disability, that problematizes the disabled body. Likewise, it is easy to conceive of the real body as the ideal/normal and the virtual body as somehow a lesser or substitute version of the real. This type of dichotomous thinking does not allow for the carnality of virtual embodiment, or the concept of presence that is clearly experienced by some SL users.

As new media technologies continue to become part of our daily routines and experiences, and as our knowledge about them expands, it is imperative to revisit terms such as online and offline, and real and virtual, particularly with regard to the notion of a body/self. Cyber worlds like SL demonstrate the complicated relationship between bodies and media, but also exist as spaces in which hegemonic norms and ideals are both upheld and transgressed. In this regard, SL bodies can tell us a great deal about what it means to both be and have a body in the twenty-first century.

NOTES

1. Victor Philip Victor, "Virtual Affair Ends in Real-Life Divorce," ABC News International, November 14, 2008, at http://abcnews.go.com/International/SmallBiz/Story?id=6255277&page=1
2. Jeremy Olshan, "Geeks Divorce After Virtual Affair," *New York Post*, Friday, November 14, 2008, 3.
3. Donna Haraway, *Simians, Cyborgs and Women: The Reinvention of Nature* (New York: Doubleyday, 1991).
4. Frank Biocca, "The Cyborg's Dilemma: Progressive Embodiment in Virtual Environments," *Journal of Computer-Mediated Communication* 3, no. 2 (1997); available online at http://www.ascusc.org/jcmc/vol3/issue2/biocca2.html
5. Michael Rymaszewski and Wagner James Au, eds., *Second Life: The Official Guide*, 2nd ed. (Indianapolis: Wiley, 2008), 4.
6. Mark Stephen Meadows, *I Avatar: The Culture and Consequences of Having a Second Life* (Berkeley, CA: New Riders), 34.
7. See Tom Boellstorff, *Coming of Age in Second Life: An Anthropologist Explores the Virtually Human* (Princeton, NJ: Princeton University Press, 2008), 16–24, for a comprehensive discussion of terms such as virtual, actual, online, offline, cyber, and so on and the meanings, uses, and debates surrounding this dichotomous nomenclature.
8. Kristine L. Nowak and Christian Rauh, "The Influence of the Avatar on Online Perceptions of Anthropomorphism, Androgyny, Credibility, Homophily, and Attraction," *Journal of Computer-Mediated Communication* 11, no. 1 (2005); available online at http://jcmc.indiana.edu/vol11/issue1/nowak.html
9. Ibid.
10. Dennis Waskul, "Ethics of Online Research," http://venus.soci.niu.edu/~jthomas/ethics/tis/go.dennis
11. Malin Sveningsson Elm, "How Do Various Notions of Privacy Influence Decisions in Qualitative Internet Research," *Internet Inquiry: Conversations*

about Method, eds. Annette Markham and Nancy K. Brown (Thousand Oaks, CA: Sage 2009), 77.

12. Christine Hine, *Virtual Methods: Issues in Social Research on the Internet* (New York: Oxford, 2005), 18.

13. For example see Christine Hine, *Virtual Ethnography* (Thousand Oaks, CA: Sage, 2000); Hine, *Virtual Methods*; Boellstorff, *Coming of Age in Second Life*; Radhika Gajjala, "Response to Shani Orgad," *Internet Inquiry: Conversations about Method*, eds. Annette Markham and Nancy K. Brown (Thousand Oaks, CA: Sage, 2009), 61–67.

14. For example see T. L. Taylor, *Play between Worlds: Exploring the Online Gaming Culture* (Cambridge, MA: MIT Press, 2006); Lori Kendall, *Hanging out in the Virtual Pub: Masculinities and Relationships Online* (Berkeley: University of California Press, 2002); Daniel Miller and Don Slater, *The Internet: An Ethnographic Approach* (Oxford: Berg, 2000).

15. Mike Featherstone, "The Body in Consumer Culture," *Theory, Culture and Society* 1 (1982): 18–33; Anthony Giddens, *Modernity and Self-Identity: Self and Society in the Late Modern Age* (Stanford: Stanford University Press, 1991).

16. Pierre Bourdieu, *Distinction: A Social Critique of the Judgment of Taste* (London: Routledge, 1984).

17. Frank Biocca and Kristine Nowak, "Plugging Your Body into the Telecommunication System: Mediated embodiment, Media Interfaces, and Social Virtual Environments," in *Communication Technology and Society: Audience Adoption and Uses* eds. David Atkin and Carolyn Lin (Cresskill, NJ: Hampton Press, 2002), 407–447; Sherry Taylor, *Play between Worlds: Exploring the Online Gaming Culture* (MIT Press: 2006).

18. Anne Balsamo, "Forms of Technological Embodiment: Reading the Body in Contemporary Culture, in *Cyberspace/Cyberbodies/Cyberpunk: Cultures of Technological Embodiment*, eds. Mike Feathestone and Richard Burrows (Thousand Oaks, CA: Sage: 1995), 215–237: Biocca and Nowak, "Plugging Your Body into the Telecommunication System," 407–447.

19. Frank Biocca, "The Cyborg's Dilemma: Progressive Embodiment in Virtual Environments," *Journal of Computer-Mediated Communication* 3, no. 2 (1997); available online at http://www.ascusc.org/jcmc/vol3/issue2/biocca2.html

20. Taylor, *Play between Worlds*; Boellstorff, *Coming of Age in Second Life*; Paul Dourish, *Where the Action IS: The Foundations of Embodied Interaction* (Princeton, NJ: Princeton University Press, 2001).

21. Steve Woolgar, ed. *Virtual Society? Technology, Cyberbole, Reality* (Oxford: Oxford University Press, 2002).

22. Ibid.

23. Irving Goffman, *The Presentation of Self in Everyday Life* (New York: Doubleday, 1959).

24. Rymaszewski and James Au, eds., *Second Life*, 10.

25. Ibid.

26. Chris Shilling, *The Body and Social Theory*, 2nd ed (London: Sage, 1993).

27. Maurice Merleau-Ponty, *Phenomenology of Perception* (London: Routledge, 1962).

28. Avatar's bodies do not completely mirror fleshy bodies in terms of leaking fluids such as blood, phlegm, urine, and so on. They are also not susceptible to injury caused by a puncturing of the skin or the demands of everyday evacuation and bodily maintenance. For example, they don't have to change a virtual tampon or diaper, as the case may be. Notwithstanding, there are bathrooms in SL and I know of one avatar who is able to shoot cum. As SL technologies advance, avatar bodies may more closely resemble our own open, permeable, messy bodies.

29. Simon Williams and Gillian Bendelow, *The Lived Body: Sociological Themes, Embodied Issues* (London: Routledge: 1998).

30. Lennard Davis, "Constructing Normalcy: The Bell Curve, the Novel and the Invention of the Disabled Body in the Nineteenth Century," in *The Disability Studies Reader* ed. Lennard Davis (London: Routledge, 1992), 9–28.

CHAPTER 11

TRAUMA'S ESSENTIAL BODIES

MAURICE E. STEVENS

USABLE THEORIES OF THE BODY AND EMBODIMENT HAVE PROVEN ESSENTIAL, and even elemental, to the emergence of Trauma Studies in all of its camped and contested manifestations. For in every instance out of which a theory of trauma has arisen—a theory, that is, describing or accounting for the disruption of experience and its representation, the rupturing of the subject's capacity to regulate its own sense of embodiment—there has also been a substrate that functions to record the suffering or rupture that fuels the engine of Trauma Studies; a medium in or through which traces of the event, real or imagined, find their expression. Trauma Studies registers these traces, appearing as they do in the form of scars, or symptoms, or lapses, or repetitions upon "bodies" of various types: corporeal entities, psychic projections, narratives of selfhood, and informational archives. Indeed, beginning with "marks" like bloodied bodies, ruptured minds, incomplete narratives, or riddled archives, Trauma Studies provides explanatory narratives that, by offering one telling of how the subject achieved its ruination, support fantasies of an originary time before the fall; a time of whole, coherent, innocent selfhood and uncorrupted, clean and proper subjectivity. And now, as much as at any other time, perhaps, the proliferation of affect and informational economies, the society of control and surveillance thus occasioned, and the precariousness of existence inherent to this stage of globalization, almost necessitate a very powerful recuperative fantasy. What better time for the ascendancy of a way of knowing injury that presents individual, whole, biologically coherent bodies as its objects of analysis; discrete and spectacular injuries that can be identified and healed, while at the same time providing technologies for managing large pools of affect and the populations understood to be defined by them.

Trauma Studies, through its tender and unfaltering facing of the sac-
rificial horrors of disintegration, promises ritualized healing, recuperative
imagining, and our admission to the global community of the healed, the
cured, and the normative. Trauma Studies and its organizing myth can only
make these promises, however, after the assumption of significant injury
has been accepted, and a recognizable sign of damage proffered. Whether
it is the body as the autopoietic self-sustaining system of Freud's imagining whose "protective shield" is breached; or the organismic body whose
capacity to experience or signify, Cathy Caruth and others suggest, is simply
overwhelmed by the immensity of the "traumatic" event; within the terrain
determined by and articulated through the emergence of Trauma Studies,
imperfect bodies and examples of hobbled embodiment, do this work, by
providing the constitutive outside to our dreams and trajectories of wholeness and control; an outside continually reconfirmed by our masterful piec-
ing together of its fragmentation.

As it is popularly, and often clinically configured, "trauma" describes
events unique in their ability to disrupt or confound both what we believe
we know about bodies (our own and those of others), and our ways of
knowing about embodiment. In most renderings, trauma's power is dae-
monic, it is scourge to ontology *and* epistemology, "unmaking worlds"
and disorienting time/space. Trauma viewed from these heights is the site
of the Dionysian rupture that signals the Apollonian return of order, the
overturning of earth that promises new crops. With the proper application
of technique and the perseverance of survivors, integration, orientation,
and ordered sociality will return. This vision of trauma—our popular and
clinical imagining of the concept—is heroic. At the same time that overwhelming events split the past off from the present by unmaking rituals
of the familiar that promise timelessness, the concept of "trauma" and its
theorizations suggest a futurity. Bringing always the undoing, and always
the promise, the daemon "trauma" names is neither evil nor beneficent,
but simply a force of upheaval. Trauma Studies promises to bring the overwhelming, the numinous, the maddening, and the ruining, under control
and into corporeal management by providing a narrative of/for understanding. That bodies are both the site and stake of the struggles for explanation
Trauma Studies champions should perhaps come as no surprise, given that
discursive practices that depend upon and reproduce concepts of trauma
have historically been critically restricted by classifications of difference.
While *harm* may be a universally applicable notion, exempt in the abstract
from the fetters of politics or prejudice, *having been harmed* is quite another
matter, an assessment whose possibility is conditioned by cultural context.
Having been harmed is as much a situated knowledge and experience, as
knowing one's injury as "trauma."

Examining the place of bodies and embodiment, and considering how notions of materiality, corporeality, and interiority condition Trauma Studies' various formations, is an interdisciplinary undertaking. This is true not only because those who engage in the study of trauma can be found across disciplines, which is to say, because trauma studies are multidisciplinary; but also, and more importantly, because thinking and producing knowledge about that which challenges ontology and epistemology, almost by definition, must be an interdisciplinary engagement. Like the notions of embodiment trauma theory requires, its connective tissue, or the oceans connecting its islands, theories about the social and about subjectivity, are similarly connected by their fault-lines, by their failures and fantasies of coherent boundaries. Indeed, thinking about bodies and embodiment in trauma studies requires drawing from and contributing to both humanities and social sciences scholarship as such thinking can only take shape in the tensions among clinical service provision, medical anthropology, cultural studies, gender studies, psychoanalysis, critical legal studies, critical race theory, performance studies, sociology, and literature.

WHAT TRAUMA MAKES OF THE BODY: OR WHY CRITICAL TRAUMA THEORY?

As a concept, trauma has been around in one form or another since the late nineteenth century and from the start, its meanings, subject to ideological and fiduciary struggle, have shifted and transformed. As one might expect, trauma has also been racialized, sexualized, gendered, and classed from its inception. In fact, from its first applications in the explanation of symptoms deriving from railway accidents, trauma has really never functioned transparently or equitably and has never been an unencumbered descriptive term. For as soon as victims began making claims on their injuries, as soon, that is, as the *harm* attending this particular form of industrial movement had its place in the lexicon of litigation, insurance agents working in the service of railway companies, and the surgeons in their employ, began defining who could and who could not be understood as having been traumatized and just what the nature of apparently nonsomatic injury might be. These were *scientific* determinations that fell then, as they do now, along axes marked by *cultural* categories of social differentiation; and that rose, as they often do, buoyed on the thermals of emergent technologies of medical detection and medicolegal reasoning. At stake in these early conversations was the question of whether an actual injury to the spine had occurred, whether notions of neurasthenia and hysteria were better rubrics for framing symptoms, or whether claimants against the railway companies were merely "malingerers" whose claims, while reflecting growing anticorporate sentiment and anxieties about industrialization, were simply fraudulent.

Between 1866 and 1890 doctors debated the significance of "nervous" symptoms, which resulted from railway accidents in which little or no physical damage was evident. What began as strictly somatic explanations for the symptoms were, after much debate and litigation, replaced with interpretations that understood the symptoms to be psychosomatic, resulting ultimately in the emergence of notions of traumatic neurosis that have remained at the core of contemporary understandings of trauma.

English physician John Eric Erichson was the first researcher to actually constitute a syndrome to explain the presence of emotional effects of traumatic events.[1] Although his claim that symptoms like the "revulsion of feelings" or being possessed of a "vague sense of alarm" and inability to "attend to business" were strictly somatic and related directly to "compression of the spine" could not be verified through the medical technologies available to him, it was, at first, popularly accepted. Indeed, his work *On Railway and Other Injuries of the Nervous System* became the primary source of medical expertise used by lawyers representing litigants against railway companies in the decade following its 1867 release.[2] The early efficacy of the somatic interpretation of symptoms emerging in the aftermath of railway accidents and visited upon railway passengers, can be explained, in part, by the particular kind of "body" central to the operations of English and American tort law, the legal setting within which claims of injury and negligence were litigated. In the context of tort law at the turn of the century and continuing through today, the subject before the bench possessed a *machinic* body whose ability to labor, or not, supplied the grounds for assessing damages. The body central to Erichsen's figuring of the postcatastrophe subject was one exhibiting the effects of undetectable, but nevertheless present, inflammation of the membranes surrounding the spinal cord and the cord itself.[3]

While it was motivated by a desire to limit the volume of successful litigation against railway companies and to constrain the scope of what kind of injury and which injured bodies could be understood as actually having lesions of the spinal tissues and cord, the work of the National Association of Railway Surgeons and the research they supported into and through the 1870s, ultimately functioned to privilege psychosomatic explanations that allowed for the notion that fear and/or alarm could produce a bona fide disorder initially referred to as "nervous shock," that was understood to be the psychologically induced "mimicry" of damage to the nervous system.[4] What was significant and new about this shift was its claim that nervous shock (something induced by fear and not "actual" injury) could itself produce physical manifestations in the nervous system in something like the "volunteerism" of hypnosis, much in the same way that Charcot had been framing hysteria during this period (indeed Charcot would later claim as much in his own work[5]). Thus, the body of trauma went from the somatic object that once damaged would likely not be repaired to the malleable body that could be a more suitable subject of recuperative fantasy.

In the context of clinical service provision, Posttraumatic Stress Disorder (PTSD) has, perhaps most effectively, instantiated the notion of the body as text from which harm can be read. PTSD, the diagnostic category used to describe symptomatic responses to trauma in relation to mental health, and the clinical object that ascribes evidentiary value to the idea that an event *actually took place* even when physical traces are not legible, has itself existed as a distinct clinical disorder for more than forty years and has seen the development of an extensive body of research and multiple clusters of investigation grow up around it. There are multiple professional societies and journals committed to the exploration and understanding of PTSD, for example, and while its hegemony is not complete in clinical settings, it provides what Ian Hacking might call a theoretical taxonomy under which newer diagnostic categories have become possible. Literally thousands of scholarly and professional articles have been written on the topic and hundreds of symposia dedicated to discussing trauma and PTSD in disciplinary contexts from literature to social work, ethnic studies to psychiatry. Indeed, in many contexts the two concepts have become conflated in ways that shape their mutual functioning. Centering fixed notions of the body as text-to-be-read in a "realist" way, this research has both provided frameworks that allow us to operate with very specific definitions of trauma and simultaneously presented a universal notion of trauma purporting to describe a very broad range of experiences. Indeed the conflation of PTSD with trauma has also imposed very fixed logics of the body (what I elsewhere call *photobiologics*) onto notions of embodiment, such that we see them repeated across other institutions of social engagement and management.

Like many ideas having their roots in psychology and medicine that have made their way into popular exchange, we find ourselves using the language of trauma easily; and often with a very powerful and felt sense that we know what we mean when we do so. The use of this nomenclature also performs cultural work by identifying those of us who use it as psychologically savvy, as empathetic and modern sentimental subjects. Surrounded as we are by mediated uses of signifiers like "trauma," "traumatic," "traumatizing" and "PTSD," we have come to learn that they relate to experiences that traverse the spectrum from simply anxiety-provoking to psychically overwhelming, or from merely physically trying to threatening bodily integrity. Although these terms appear to have become generally evacuated of their specific meanings, most of us believe, at base and instantly, that we know exactly what to look for when cued by these troubling signs.

The logics of trauma state that while its immensity renders its representation impossible, its traces in the form of symptoms inscribed on the body or in anomalies of the signification of embodiment, guarantee that an event actually took place. The symptoms we can observe and tick off our diagnostic list, the scars we can delicately trace with exploring eyes or fingers, the landscape whose barren patches we can lament, these are all signs in the here/now that assure the facticity of an event in the there/then. In each of these logic schemes, there is a two-dimensional representation (e.g., the "blackskin" epidermal screen, the present moment of symptom presentation, and the "skin of the photo") that stands in as a representative of the real and absent signified, both evidence and ghost.

Neither indexical nor symbolic signifier, trauma has taken on the logics of the icon. When we imagine we are "seeing" trauma or the signs of its passage, we know immediately that something spectacular and catastrophic has transpired and we fear, also with a sense of immediacy, that normal systems for understanding the event and any of its survivors will be overwhelmed and rendered incapable of adequately capturing its immensity or the subtlety of its sublime pervasiveness.[6]

However, the simultaneous sense of "knowing" something has transpired, and the utter frustration of having our understanding overcome by trauma—of not being able to render that experience legible through representation—has made its clinical and theoretical application particularly vulnerable to social emplotment imbedded in the concept of "trauma" itself. Trauma, as a situated knowledge that emerges from the specificities of the moment in which it is invoked as an appropriate or obvious label, bears in rather remarkable ways, traces that reveal its cultural work. This level of vulnerability and its ramifications poses a central point of concern in relation to our understanding of the body and embodiment in as much as racialization, sexualization, and the tyranny of the visual shape what trauma can be, which subjects its signification hails, and which institutional practices it underwrites because they are understood as adequate to its amelioration. I am reminded of a colleague who tells the story of providing mental health related services to an African American man of middle age who although exhibiting symptoms of PTSD was unable to receive the diagnosis (and subsequent disability remuneration) because the only "cause" he could provide, the only instigating event he could conjure, he took from the introduction to James Baldwin's *Evidence of Things Unseen;* a book wherein Baldwin recounts his investigation into the Atlanta Child Murders.[7] My colleague's client could say only that he had been deeply affected and disrupted by the everyday cost of "growing up a black boy in a white country." The fact that my colleague could close this narrative by describing the "fortunate" event of her client having been unexpectedly struck by a car, and thereby having obtained a "precipitating event" that could be used to justify the diagnosis of PTSD and the subsequent successful acquisition of governmental support, says something of the kinds of subjectivity and experience that become legible before the bench of popular ideas about trauma and those who can be traumatized. From here, sites of invisible injury proliferate along lines of flight newly imagined: whales who beach following navel tests of new sonar subsonics, agricultural economies infiltrated by genetic adjustments to seed (not to mention their dramatically suiciding tenders), the thirsty survivors of Katrina, the unrepatriated bones of tribal ancestors, the social networks of elephants shattered by violently acting out groups of adolescent males. Indeed, notions of trauma emerge as often very complex "sets of practice" in several cultural institutions in which the figure of the body and practices of embodiment shape cultural practice; namely, the clinic, legal discourse, cyberspace, popular culture, and, of course, the street.

At stake in my concern that the concept of trauma developed around injury related to railway accidents, wartime wounding, or overwhelming natural catastrophe, and not other sites of less spectacular or less-valued injury, is the centrality classifying systems have had in the formation of ideas about whose sensibilities or fragile connections can be disturbed by near-death experiences, whose civility can be upset by the horrific, and who can be overwhelmed by fear; who or what, in short, can be traumatized. Indeed, I concur with the increasing number of theorists growing critical of trauma, who have been arguing that many social actors and networks are inadequately understood within, or overdetermined by, its boundaries. For example, psychoanalysts might argue against the application of trauma theory in cultural study because of its misappropriation of Freud or Janet's ideas about the mimetic or antimimetic nature of traumatic memory; or ethnic or cultural studies theorists may take trauma theory to task for its inability to recognize traumatogenic institutions such as enslavement, genocidal cultural contact, environmental havoc, or the simple ubiquity of nonspectacular racial violence and microaggressions whose bodily marks are illegible behind the veils of perceived cultural difference. Transnational critics might decry the European and American impulse to force diverse peoples into the culturally specific rubric of trauma, casting aside the authority of local knowledges. For example, South African clinicians trained in Trauma Resilience or PTSD trauma response models report that the effectiveness of these approaches, based on a hyperindividualized model of the subject, produce negative response among clients whose culturally specific senses of self are both individual *and* collective.

These are all important and truly useful critiques, to which any serious consideration of trauma theory must respond. However, the concept of trauma itself must be interrogated. We must submit it to an analysis that highlights the work it does to define who can and cannot be understood as injured, who possesses the capacity for being the object of empathy, what

scales of injury warrant analysis, and whose recuperation merits the marshalling of state or global resources and research energies. Moreover, the
centrality of the body in trauma studies is also troubling. Figuring the body
as organismal and independent, as the autopoeitic closed system that sends
tendrils of desire into an othered world of providing or withholding objects,
also fixes the scale at which we can register the injury. This way of thinking
the body and embodiment also delimits the site and scale from which we
might imagine strategies or ways of knowing, of coming together to ameliorate the features of suffering that are simply debilitating of individuals,
communities, and systems of relation. These are all questions, in the main,
of embodiment, questions about the body.

TRAUMA: FROM WHAT IT "DESCRIBES" TO THE BODIES IT "MAKES"

Like most examples of "socially constructed" objects of knowledge, trauma's
force can be measured in the material effects it produces in social relations,
institutional practices, and public policy. Although critics have called attention to the limitations of trauma theory, their criticism has been primarily
academic and has not closely examined how these limitations prove problematic in specific institutional locations.

As a concept formed out of injury related to railway accidents, wartime
wounding, or overwhelming natural catastrophe, notions of class, race, age,
gender, and sex have all been central to the formation of popular ideas about
whose sensibilities can be disturbed by near-death experiences, whose civility can be upset by the horrific, whose constitutions are resilient, and who
can be overwhelmed by fear. And as an increasing number of theorists growing critical of trauma (as it is traditionally figured) have been arguing, not
all social actors are adequately understood within its boundaries. Trauma
does not simply describe subjects and/or their embodied experiences, it also,
and perhaps more accurately, *creates* them.

At the same time that increasingly rigidly defined parameters have defined
its technical (and institutionally legible) boundaries, the idea that trauma is
somehow *universal* seems ubiquitous. Its daily use to describe a very wide
range of experiences confirms that trauma is flexible and adroit, passing
from one context of expertise to another, slipping across borders to be readily
recruited to new discourses, new practices, and new contexts of explanation.
On one hand, the ability to pinpoint the traumatic event or symptom with
spatial and temporal coordinates (necessarily past and completed) makes it
particularly powerful in the clinical or diagnostic setting. There is an agent
and victim of injury, a place and time of occurrence, a physical or psychological indication of damage, and a blooming narrative of accountability or

innocence. The traumatic event possesses specificity. On the other hand, its unknowability, how it eludes signification and description, confounding our technologies for knowing and mastering it, makes it particularly susceptible to becoming something else as well. Trauma is also enigmatic.

Thus, we have a dilemma. Trauma is both specific *and* enigmatic, both discursive *and* material, both readily apparent *and* elusive. Similarly, the broad set of neurobiological responses to traumatic events (the psychophysiological responses that seem ubiquitous and reason for universalized treatment response) and the multiple variations in the phenomenological or expressive response to trauma across groups defined in terms of gender, race, ethnicity, class, and even sexuality, also obtain a tension. While we may all develop "startle" responses in the aftermath of trauma, for example, the intensity of those responses can be shown to vary dramatically in correspondence to differences in one's cultural or social positioning.

The basically arbitrary and, in some ways, theoretically counterintuitive requirement that the traumatic event have specific spatial and temporal coordinates, has primarily to do with the fact that limits to its application typically emerge in relation to *where* or *when* the "trauma" actually emerged. For example, the location of trauma's origin can make it inaccessible to the PTSD model. This is apparent in the case of acute traumatic episodes originating in sociocultural structures where the traumatogenic agent is not readily discernable. Critical race and critical legal theorists in the United States and Europe have usefully analyzed the specific damages produced in relation to the law, prison industry and immigration policy, for example. Likewise the case of trauma that exceeds individual bodily experience is also difficult to localize and thereby normalize. Categories like ongoing or repeated trauma, multigenerational institutional relations, or even the sense of *impending* trauma that can produce PTSD symptoms, are all types of trauma that fall outside temporal parameters of conventionally applied models of injury.

Rather than thinking of trauma as an identifiable and discrete event that must have occurred at some specific point in time and place, it can be framed as a cultural object whose meanings far exceed the boundaries of any particular shock or disruption; rather than being restricted by the common sense ideas we possess that allow us to think of trauma as authentic evidence of something "having happened there," a snapshot whose silver plate and photon are analogues to the psyche and impressions fixed in embodied symptoms, the real force of trauma flowers in disparate and unexpected sites of production. "Trauma" circulates among various social contexts that give it differing meanings and coproduce its multiple social effects, and its component memes, those pivotal conceptualizations that tailor its function, have origins that can be traced to coordinates that vary in time, space, and semiosis; coordinates whose ideological concerns come to refract or anchor trauma's meanings in very fixed notions of the body and our sense of embodiment. More than *describing*, the central work of trauma is that of *making*.

HISTORY AND MEMORY: A TALE FOR TIMES OF TRAUMA

Like trauma and memory itself, the *study* of memory and the formation of the memory sciences have a milieu and have taken their shape and cue from social contexts that, over the course of modern industrialization's inexorable cultural speedup, have come to privilege the production of history over the production of memory. Spaces of history like the archive, the memorial, or the "official story" are often *figured* in binary opposition to spaces considered the purview of memory: the performance, the repertoire, or the ephemera of public culture and spaces. Moreover, through the rhetorics of provenance, authenticity, and the originality of the record, institutions that manage memory increasingly wear the robes of truth's arbiters. Repositories of facts, conglomerates of evidence, memory management takes place while historicity is conferred by the archive and through its objects. Authority to produce "true" narratives shifts, as remembering bodies are replaced by bodies of information that constitute history. While they are *posed* in opposition, both memory and history contribute to a regime of remembrance whose logics and functions are familiar and, in some ways, comforting. Its logics are those of the photograph or the gene or the eyewitness testimony; its functions converge to convey truth, to represent the real and to reproduce the Same. Thus, one need not accept the opposition between history and memory to appreciate the effects produced by the solidification of their polar relation. History posed against memory works. It works like science against culture or data against interpretation, its cultural work deriving not simply from their binary opposition, but from the meanings ascribed to those oppositions and the material relations those meanings justify, the ideology they reproduce, and the incommensurability they convey.

The science of memory has shifted from conceiving of its object, *memory*, as an evolving entity open to processes of contestation, reframing, appropriation, diffraction, or simple dissolution, and has moved, again, with seeming inexorability, toward a focus on *history* as the fraught and always problematic recording of what has "gone on," as the recitation of actions and events contained within the past-perfect grammar of description, and the body's sometimes inaccurate keeping of the record. There and then was an event, it occurred in a place and at a time that are, by definition, distanced from here, from now; and the historian heroically does the work of salvage, approaching the event through documents, artifacts, and corroborating testimony believed to shed evidentiary light on the always-already past event, to link it through an ideal provenance of its traces to the present. The historian's labor and the measure of their agency or ability lies in getting as "close" to the event as possible and determining what should be memorialized in objects of historical inscription. Unlike history, as the story goes, memory exists continually, inscribed in the ongoing production of a narrativized self or community of practice or affiliation. The muscle remembers, the space is haunted, the landscape is scarred, always, with memory, a trace remains. A trace remains, defiantly, sometimes hinting, sometimes pressing, sometimes roaring, but always insisting in its ubiquitous return. History, which requires sifting through remnants instead of traces, speaks the past differently. Events captured in history are located in the mythos of temporal progressions, in the relative relation between moments and events; the distance imagined between here/now and there/then is history's necessary condition. Indeed, history, as a trope with rhetorical force, is memory's nemesis it would seem, pushing it ever flatter, out of the flesh of bodies, gestures, objects and spaces, and into the amber of dominant signs and symbols, or the architecture of archives, or the ash and bones carefully catalogued there—but not in the bodies whose experience it purports to record. In history, the *past* becomes an imaginary occupant of the symbolic, and provenance becomes its genomic real.

And yet history grows gaunt and distracted in its confrontation with events that test its ability to represent, to inscribe with any accuracy, at all. Hunched over, squinting, and losing its flexibility, it worries at the frayed ends of incomplete narratives and hidden transcripts. Still, we see that when and where history struggles, when and where it collapses in the face of the body's absolute truth of *having* pain or being harmed, and the inexorable suspicion that accompanies *documenting* it, we see, in fact, the memory *sciences* providing support. The variously institutionalized science of memory smoothes over history's lacunae, its impotencies, by disavowing the possibility that specifically racialized, gendered, sexed, and classed violence endured by particular communities might also be understood within the rubric of trauma; or that the rubric of trauma may secrete within it the necessary logics of race, gender, sexuality, or class.

Just as the invisible genome vouches for the validity of phenotype, or the effaced technologies of the photo argue the "fact" of its real representation, the past and completed unrepresentable trauma supports claims about the coherent subject of history and its occupation of a body/palimpsest. It says, "You see, there once was a whole, seamless and modern subject/body. Our effort to repair it, by making legible its injury, is proof enough of its having been there at one time, whole (read: vulnerable), pure (read: violable) and mature. Trauma has rendered this particular example of proper subjectivity and able-bodiedness damaged, where once, in a moment of innocent possibility, it was not." Of course, the wholeness, purity, and propriety of this subject, this "clean and proper body," have been built on the very particular ways it has always-already been classed, gendered, sexed, and, to focus on a particular example, raced.

RACE: THE REPUDIATED MOTE

Through its enigmatic signification, race has played a pivotal role in the formation of contemporary notions of memory, identity, embodiment, and trauma that are based on interior experiences of overwhelming exterior events. From Freud,[8] Darwin,[9] and the scientific racisms of the nineteenth and twentieth centuries, to the postpleasurable traumas of World War II and the recuperative practices of American clinical psychology and neurobiology, psychoanalytic theories and psychotherapeutic practices have been unable to take up racialization as a social process that produces some subjects as vulnerable to traumatogenic injury, and others as not. Indeed, the "Others" to this village of the traumatizable, because they are the ultimate source of phobia and, therefore, cannot be overwhelmed by it, are not imagined as possessing the psychic interiority necessary for identification and institutional legibility. Indeed, as phobic object, the Other portends both the need and possibility for cathexis. Ironically, the racialization of these others both produces and is reinscribed by the fact that the subject of psychoanalysis and recuperative treatment remains a deracialized, thoroughly modern subject, imagined through *universal* (read: identical) mechanisms and structures understood to work within *particular* psyches. In this way, the Other stands in as the constitutive outside that vouches for the uniformity of a self that possesses an unconscious composed of properly repressed drives and a social presentation replete with appropriately sublimated libidinal urges.

This period also saw the emergence of a widespread acceptance of biologistic notions of race and difference buttressed in the United States and Europe with scientific theories and epistemologies informed, at base, by a notion of incommensurable difference. This incommensurability or failure of recognition derived from and reproduced racial logics that found easy expression through the visual technologies associated with eugenics, criminology, and psychoanalysis. As a result, the convergence of Social Darwinism, for example, emerging photographic technologies, and a fledgling psychoanalysis naturalized ideas of racialized peoples as lacking the psychic interiority that could make psychic trauma, or even basic suffering, a social possibility. This is particularly significant because following Erichsen's early work with "railway spine," theorists of nonphysical "hysterical" trauma like Charcot, Janet, and Freud were building their paradigms on these epistemologies of difference. As a result, the taxonomies they developed, because informed by racially embodied notions of the other and the self, reproduced these formations in their work; these were intellectual formations along which the memory/history binary was also mapped. Ultimately, the convergence of these ideas and their inherent logics conspired to exclude the experiences of racialized ethnic communities from the category of catastrophe that could be called traumatogenic, the typology of experience that could be called history, and from the practices of its collection and discipline necessary to narrating and archiving the nation. And even more importantly, racialization provided the intersection between the "immaterial" psyche's wound and the value of the body as site of evidence.

Because the traumatized subject has been one constructed through medical, psychological, legal, academic, and cultural institutions that are themselves based on racially unmarked subjects/bodies, it makes sense to understand both the subject of trauma and trauma itself to be similarly unmarked, and framed in terms of a body that is "essentially" white. The question is, not what but *how* does this marking mean in spatio-temporal-semiotic locations that produce constellations of practice like PTSD and its enabling agents (clinics, clinicians, psychotropics, therapies, institutional recognitions, bodily performances, etc.)? If we accept that PTSD is a bundle of social practices that reflect how trauma is invoked in clinical/medical institutions, and that institutional formation produces legible subjects whose bodies will act in predictable ways—that is, he or she who has been traumatized and is exhibiting symptoms that warrant the diagnostic categorization of PTSD and the disciplinary practices that spring into action in the application of the diagnosis—then the *what* and *how* of this marking's meaning is reflected in the subjectivity/sense of embodiment produced by the diagnosis. The injured/traumatized subject is both the constitutive inside and outside through which all proper citizen-subjects can know themselves: whole, coherent, seamless, healed, and modern. These are the ephemeral traces to which we must attend, these ideal imaginings of ourselves as whole, wounded, or mended.

The enigmatic signifier, Laplanche tells us, wishes to be translated.[10] That is, its *significance* is driven by the desire to be exposed, refashioned, and represented. Because its consideration of representability is constrained by culture, its signifying path always already provoked by the classifying systems that order the differences through which its legibility emerges, because the systems of classification already possess a symbolic valence and are already related one to another; because of these factors the enigmatic signifier speaks in names that are familiar: gender, sexuality, class, and race. While reconfiguring our understanding of trauma and the logics that inform memory cannot remove the repudiated mote from the eye of the memory sciences, that which remains its enigmatic yet powerful metaphoriser; a trauma differently understood, and a memory whose racial logics are acknowledged can certainly render its material effects transparent and its representative bodies constructed, even if its signification remains opaque.

FROM TRAUMA'S NECESSARY BODIES TO...

Examining institutions of practice like clinical service provision, legal language and action, cyberspace memorializing, and popular media representations of terrorism and catastrophe can illumine what it means that experiences of trauma, diagnoses of PTSD, easy memorializing, bodily instruction, and even legal framings of unacceptable harm *make bodies*. Beyond that, though, undertaking these reflections can show how the work of trauma in one institutional location feeds into and draws upon its iterations in other institutions. How, for example, legal definitions of the tortured body rely on limiting concepts of physical and mental traumatic injury, which in turn, supply the logics and just cause to training institutions, cyberspatial sites of memorialization, and representations of terrorism and its effects. We can examine the links between contemporary representations of terrorism and the temporality of trauma and suggest that even the democratizing of suffering that contemporary terrorism discourse offers might function to ameliorate the requirement that traumatic events be restricted to a spatially and temporally distant location, or to bodies configured in fixed or limited ways. Indeed, if we imagine that rather than mere legal categories, the peculiar legal objects *hate crime* and *genocide* in domestic and international law, are actually complicated sets of practice that reflect struggles over the status of the legal subject and the body in the context of *harm*, we can come to take seriously the injury inherent to rupturing a subject's capacity to regulate their own sense of embodiment.

This possibility might find its most recent and pressing application in the jurisprudential resurrection of the tortured body and the practices that this (and the US engagement in empire building) has emboldened. In addition to exploring traumatic iconography and representations of terrorism, torture-related jurisprudence and contestations over the definition of genocide as sets of practice that exceed the parameters we might normally expect in investigations of the law or the media suggest new meanings for the body and embodiment. On another front, we can extend our notions of the body and embodiment by analyzing trauma's manifestation in clinical settings by focusing on PTSD not simply as a diagnosis, but as a set of practices that include service utilization, diagnosis, psychotropic medicating, imaging technologies, hospitalization, and efforts to revise the Diagnostic and Statistical Manual. Limiting conceptions of trauma have shaped the basic assumptions and material activities attending notions of harm, injury, and their subjects. Shifting from our conception of trauma as a descriptive term, and moving to thinking of it as a concept that makes subjects and shapes bodies through the function of significant social institutions, can help us determine and propose alternative approaches to assessing and responding to our social suffering without recourse to disavowal.

CONTROL SOCIETIES AND TRAUMA AS APPARATUS OF THE BIOPOLITICAL

Catastrophe, and the "trauma" we imagine it to convey, that is, the sense we make of the injury and suffering inherent to catastrophe's upheavals, is quite central to forms of imperialism and global neoliberalism that seem to be coming so fully into their own over the past decade. That is, what sentimentalism was to the imperial and colonial projects of the last century, where laboring subjects were brought into modernist economies as to-be-subjectified, to-be-disciplined, citizen-subjects—what sentimentality provided this great shift, trauma now provides the control society. And those of us interested in shifting possibilities in the framing of bodies and embodiment should be concerned with this. Where instead of, or rather in addition to, disciplined citizen-subjects, we might speak of observing the emergence of controlled populations of affective flow and practice, what many have talked about as the emergence of a biopolitical economy.

Trauma as a particularly understood catastrophe calls for relief and strategies of recuperation. It acts as the point of the sword that opens lines of entry, vectors for economic flows operating under the best of intentions; operating with the function of allowing certain bodies (ours) to be our best selves, the best of humans. This activity of opening economic flows is salved with our own pleasurable sentiments. Thus, while laboring to manage affect, to improve emotional states, to alleviate suffering, it is precisely our own affect that is turned to the work of globalizing capital flows.

Indeed, information technologies, diagnostic practices, and biotechnological expertise designed to manage bodies and to territorialize notions of embodiment, follow these flows; and with them also flow the instruments of their management: archives, inventories, tables, outcome measures, and so on. The great globalizing science of *eu-affectics* then requires the ready emergence of political economies that accompany securitizing and the bringing of control, the provision of safety, and the hope of revitalization. And while the scale of what is essential about trauma's "essential bodies" has multiplied, operating now at the level of "individual" bodies and their injuries as well as at the level of the body's affects, we see, nevertheless, that "Trauma" has its bodily needs.

NOTES

1. John Eric Erichsen, *Concussion of the Spine: Nervous Shock and Other Obscure Injuries of the Nervous System* (London: Longmans, Green, 1882).
2. John Eric Erichson, *On Railway and Other Injuries of the Nervous System* (Philadelphia: Henry C. Lea pub), 1867; see also E. Brown, "Regulating Damage Claims for Emotional Injuries before the First World War," *Behavioral Sciences and the Law* 8 (1990): 421–434.
3. One cannot help but note the similarity of arguments around Posttraumatic Stress Disorder and Traumatic Brain Injury (TBI) debates currently raging around veterans of the Iraq and Afghanistan military campaigns. Indeed, the Pentagon and Veteran's Affairs researchers taking up the conversation can be placed along the spectrum between somatic and psychosomatic interpretations of symptomology manifesting among veterans. The recent struggles over the awarding of the Purple Heart and the various VA-related benefits it confers upon veterans reflect these contestations quite clearly.
4. Here again we can find resonances in contemporary debates about trauma. Ruth Leys has famously critiqued Cathy Caruth's insistence on the traumatic symptom as the "literal return" of the previously unexperienced event (*Unclaimed Experience*), as an error of mistaking what is essentially a mimetic representation for an antimimetic one. See Ruth Leys, *Trauma: A Genealogy* (Chicago: University of Chicago Press, 2000).
5. For more on this, see Charcot's *Clinical Lectures on Diseases of the Nervous System*, trans. T. Savill (London: New Sydenham Society, 1889).
6. Other similarly troublesome categories come to mind here, experiences that also often fall into the mode of iconographic signification: religious experience of the "(W)Holy Other," engagements with Jung's "numinosum," psychotic or "outside" experience, institutionalized "micro-aggressions," and so on.
7. James Baldwin, *The Evidence of Things Not Seen* (New York: Holt, 1985).
8. Sigmund Freud, *Totem and Taboo: Some Points of Agreement between the Mental Lives of Savages and Neurotics.* ed. James Strachey (New York: W.W. Norton, 1989, c1950).
9. Charles Darwin, *The Descent of Man, and Selection in Relation to Sex* (New York: Hurst, 1874).
10. Jean Laplanche, *Seduction, Translation, Drives*, eds. John Fletcher and Martin Stanton (Psychoanalytic Forum London: Institute of Contemporary Arts, 1992).

CHAPTER 12

HOLD ON! FALLING, EMBODIMENT, AND THE MATERIALITY OF OLD AGE

STEPHEN KATZ

MIKE HEPWORTH, a leading writer in cultural and critical gerontology, remarked that "if the body did not age there would literally be no gerontological story to write or read."[1] Whether such stories are literary, popular, scientific, scholarly, or fantastical, the body is the foundation for what we know, experience, and imagine the aging process to be. In Western culture the dominant story about aging bodies is framed around the biology of decline. Physical aging appears to represent the deepest, most natural and most obvious truth about what it means to age. This story has been the basis for the development of the gerontological sciences since the nineteenth century and their emphases on the aging body's risks, vulnerabilities, functions, losses, and deficits. Today, the biological aspects of aging have been articulated within fields of expertise that include gerontology and geriatrics, the life sciences, the healthcare professions, pharmacology, the neurosciences, and even antiaging medicine. However, behind the dispassionate objectivity of the scientific enterprise lies a broader cultural background of contradictory images that marginalize, denigrate, and desexualize older people, yet obligates them at the same time to resist their own aging through active and independent lifestyles. Hence, the dominance of scientific discourses about aging bodies must be seen with this background in view.

A further breakdown of the science story about physical aging is evident when we consider that dominant cultural narratives are negotiated, challenged, and subverted within our lives as part of the aging process itself. We have bodies, which age, but we also have embodied lives whereby we create subjective phenomenological dimensions of meaning and identity. As embodied subjects, we are reflexively bound to make intelligible the physical changes and passages of life through which we experience living in time. This is biographical work, the products of which are plentiful in art, religion, literature, and the everyday narratives of older individuals. Biographical aging also provides the inner resources for decision making and imaginative understanding about the dilemmas of maturity and longevity. Such resources allow us to adapt and shape our aging identities in relation to various bodily states, including frailty, disability, and decline as well as those typified as "healthy," "active," and "successful."

In this chapter, I argue for a perspective of *embodied aging* that explores the coalescence of physical and biographical aging. To do so I look to a framework on the materiality of aging bodies that eschews the universalism and determinism of dominant science narratives and the static subject-object, self-society, and mind-body dichotomies underlying traditional cultural discourses on aging. The materiality of embodied aging approaches the aging body as both creator and product of the experiences configured by our material worlds, such as the spaces we live in and environments in which we move. To elaborate this argument, the first part of the chapter reviews the relationship between aging studies and social-constructionist body studies, whose mutual neglect of aging bodies has been challenged recently by writers who emphasize the materiality of embodiment in everyday contexts and practices. The second part of the chapter applies this emphasis to the empirical example of falls, falling, and fall-prevention programs, questioning how falls embody old age physically and biographically. Bodies fall. But falls are experienced, narrated, and treated by older people in ways that can dramatically change their health status from independent to dependent, cause a transition in place from home to care institution, and shift a resilient identity to a vulnerable one. The chapter's conclusions consider the theoretical importance of falls both to body studies and aging studies. Throughout this writing I refer to the "aging body" as a conceptual category, akin to other types of "body" identified in the critical literature (e.g., the consuming body, the civilized body, and the deviant body). However, there is no singular aging body; rather there are (constructed) male and female bodies, further differentiated by race, [dis]ability, class, sexuality, and region, whose aging is shaped within social hierarchies of difference and inequality.

PART I: AGING BODIES AND BODY STUDIES

THE PRESENT/ABSENT AGING BODY

Today, as the prospects of extending human longevity have inspired some of the most challenging and haunting images of the powers of modern science (and science fiction), the preoccupation with aging bodies in the public imagination has become pervasive. However, a critically theoretical and reflexive treatment of bodies remains mostly absent in aging research. Although many gerontologists, positive-aging advocates, and antiaging activists increasingly attack conventional biomedical models for ignoring the personal, social, and environmental factors in the aging process, this group tends to abandon the body to the same biomedical models it critiques. In other words, the body is treated as somehow outside of culture, as a physical entity best left to the life sciences and the dominant science narrative.

Rather than forsaking the realities of physical aging, Kathleen Woodward advises researchers and writers not to "detach the body in decline from the meanings we attach to old age."[2] Other critical gerontologists also caution that the importance and visibility of physically aging bodies in "old, old age" or "Fourth Age" decline should not be clouded by the "positive," "active," or "successful" "young old age" or "Third Age" ideals and ways of life that dominate popular culture.[3] Indeed, the currently positive reversal about the diversity and contingencies of healthy aging in retirement has further deepened the negative aspects of aging and dependency associated with the "Fourth Age" as a kind of "metaphorical 'black hole' of ageing" where all agency, mindfulness, mobility, and selfhood are seen to collapse.[4] For these and other reasons, Peter Öberg observes that the mixed presence and absence of the aging body within the academic, professional, and lay communities constitutes a profound paradox.[5] The stirring of public anxiety about aging is such a forceful tenet of consumer culture that images of even the most minor signs of physical aging are widely circulated as threats to social success. And yet, despite the ubiquity of such images, realistic images of aging bodies are few and far between, hidden from view as a kind of geriatric pornography that might disturb postmodern cultural fantasies about timeless living.

This paradox is particularly vexing given the influence of subfields such as the sociology of the body.[6] Inspired by various structural, feminist, and phenomenological traditions developed by Bourdieu, Butler, Elias, Foucault, Merleau-Ponty, and others, sociologists have looked to the body to pursue a wider critique of consumer culture, social regulation, medicalization, gender performativity, and the "civilizing process" itself. These critiques certainly address the absence of bodies and embodiment in social research; however, they devote scant attention to the problem of the absent *aging* body and its paradoxical relationship to the sensationalization of physical aging in popular culture. The intellectual tensions between body studies and aging studies reflect the paradoxical position of older people themselves. With their bodies, even in absentia, implicated in all the contradictory cultural and moral orders with which they identify, older people can experience the embodiment of aging as a fractured process of resisting, accepting, denying, and recreating aging.

If sociocultural critiques of the body have borrowed little from gerontology and aging studies, the reverse is not necessarily the case. Theoretical gerontologists have looked to these critiques to explore embodied aging in ways that correct their field's lack of research in the area.[7] Aging body studies have also incorporated gay/lesbian,[8] performance,[9] and humanities[10] perspectives to create new critical thought spaces. However, it is the dialogue between feminists and gerontologists that has broken particularly important ground since the 1970s. Feminist gerontologists have criticized the feminist community for its neglect of age and aging bodies and encouraged both gerontologists and feminists to bridge antisexism with antiageism, and sexual inequality with age inequality.[11] Feminist research on age links the youth-based sexual politics of appearance with our culture's disparaging of older women, as is evident in the history of hormone replacement technologies and the biomedicalization of menopause.[12] For both women and men, feminist writing pinpoints that the imperative to grow older *without* the visible signs of aging pervades our media, fashion, dieting, and cosmetic industries. And yet, to grow older without aging implies unvarying functionality, permanent performance, unfailing memory, and unceasing activity. Such impossible standards affect women most prominently because of the cultural idealization of their bodies as age-defying technologies. Otherwise, as Clary Krekula comments, "Older women have predominantly been studied from a misery perspective, stressing women's ageing as a problem."[13]

The vital exchange among feminist, body, and aging studies strengthens the interdisciplinary flow between and among the arts, sciences, and professions and counteracts assumptions about the aging body that stem from dominant cultural, constructionist, and biologistic biases. It also provides a theoretical basis for understanding the *materiality* of aging bodies and what this means in the context of embodied aging.

THE MATERIALITY OF EMBODIED AGING

Allison James and Jenny Hockey claim that "hitherto the materiality of the experiencing body has been somewhat sidelined in attempts to *problematise* biological accounts of the body offered by the medical model and so progress the social constructionist agenda."[14] Such sidelining can be seen in critical and feminist literatures where the biological materiality of bodies disappears within overly deconstructive or discursive analyses. The neglect of biology in the name of antiessentialism can also neglect those limited or defined by their biology, such as older or disabled individuals. In reaction to these problems, several body theorists who critique technoscientific and biomedical practices have created alternative frameworks that blend biological and constructivist positions in nonreductionist ways. Examples are Simon Williams's critical realist perspective on health, Myra Hird's "nonlinear" materialist critique of sexual reproduction, and Chris Shilling's "corporeal realist" strategy to bond bodily and social interdependence.[15] I see this trend emerging in the aging field where body researchers such as Toni Calasanti, Julia Twigg, and Emmanuelle Tulle[16] also take the constructivist body literature to task by maintaining, as Twigg states, that the "physiological significance of the body is undeniable."[17]

What gives this trend its critical currency is its focus on the *materiality of embodied aging*; that is, the inseparability of the physical realities of aging from their lived material contexts, without succumbing to dominant biologistic narratives about the aging body. Several researchers in dementia research, borrowing from Merleau-Ponty's work on the phenomenology of the body, have advanced this perspective.[18] Caregiving, however, is the most obvious illustration of embodied aging because of the intimacy of the bodywork involved. As Julia Twigg demonstrates in her research on bathing and carework in Britain, strict scheduling and procedural routines transform the quotidian act of bathing and its physical comforts into an enactment of power relations. Older people "experience their bodies in the contexts of a profound cultural silence"[19] as the action of a (younger) clothed and gloved careworker washing the naked body of an older person can result in feelings of embarrassment and shame. And, as Twigg observes, if the carework around aging bodies is typically considered dirty and demeaning, then the care-recipient will think of their own bodies as unsightly and contaminating. Thus, for both care-recipient and careworker, in the short space of a fifteen-minute bath and its embodied microcosm of touch, sensation, discretion, and vulnerability, all the deep-seated ambivalence that characterizes cultural attitudes about aging bodies bubbles forth.

The materiality of embodied aging manifests itself in biographical aging through the inner struggle of negotiating and integrating various age-related identities through time. This inner work is akin to what Simon Biggs calls "the mature imagination,"[20] a complex point in the life course whereby the need for late-life biographical integrity is offset by our ageist society's asymmetry between inner experience and outer appearance. Today, according to Biggs, "whereas attention to appearances emphasizes inter-changeable options for the performance of identity, depth anchors the self in memory, continuity, desire, and enduring motivation."[21] The embodied expression of this identity rift between surface and depth is clearly observable in performance cultures such as sport and dance because optimal physical prowess is expected while careers end at relatively young ages.

For example, Steven Wainwright and Bryan S. Turner's ethnographic research of the Royal Ballet of London explores the powerful influence of images of youthful bodies in the world of professional classical dance.[22] Here, the dancers learn to embody and perform the style, stamina, and competence of an elite physical capital that rewards them with status and self-definition. However, the attainment of such capital requires constant discipline to manage the severe demands of rehearsal schedules, strict diets, and recovery from injuries and exhaustion. By early middle age, most dancers face the advancing constraints on their physical abilities. Retirement from their short-lived careers is difficult, however, because of the intense association of the dancer's identity with their body. Wainwright and Turner emphasize the point that the self-fashioning of the performing body inevitably creates an internal dialogue between the performers' capacity to adapt to the wear and tear on their aging bodies and the transient nature of physical capital. Their coming to a mature imagination is forced because of the extreme physical nature of the dance world.[23]

Other sports that seek to prolong rather than foreshorten athletic performance also make for fascinating study about the negotiation of physical capital within a field of material interaction between identity, biography, and embodied aging. Tulle writing on veteran Masters long-distance runners portrays an intense athletic culture where the body is the medium for expressing and understanding aging.[24] This category of runners is open to older age groups who, like younger runners, maintain their status through rigorous training, injury and pain management, and disciplined exercise and record-keeping. Tulle argues that the older runners work on their bodies *as* aging bodies, adopting specific and taxing strategies to deal with the loss of strength or the risk of injury, while being very aware that these strategies defy traditional images of aging and decline. The runners come to know their aging through the intimate interpretation of their bodies' performance, where pushing the limits defines who they are as older subjects. Thus the veteran runners negotiate their aging identities through training and performance that neither resist aging nor disembody it, because the runners "have endowed themselves with bodies that are both ageful and competent."[25]

In my view, field studies of caregiving, sport, and dance performance stretch the interdisciplinary exploration of aging bodies in two important ways. First, they demonstrate that embodiment is a materializing process whereby the vicissitudes of physical and biographical aging are grounded in bodywork practices, routines, and environments. Second, such studies show that aging folds together Self and body through activities that both sustain and disrupt identity. Age identities and the "mature imagination" are continually interpreted, narrated, and negotiated in relation to physical aging in various ways. With these points in view, the next part of this chapter treats falls and fall prevention as a field of embodied aging that illustrates a critical approach to aging bodies within gerontological knowledge.

PART II: FALLS AND EMBODIED AGING

THE GERONTOLOGY OF FALLING BODIES

> If we do not make this a topic of public discourse, if we do not get the issue of falls prevention into the conventional wisdom, if we do not make it sexy and glamorous, then we will not be prepared for the inevitable falls of more than 75 million baby boomers, who are marching to their inevitable, actuarial destiny. They're going to get old, they're going to get frail, they're going to wear down, they're going to have their share of accidents, and then the full economic and medical and social and health implications of what it can do will certainly be on the front page of *USA Today*.[26]

This gloomy scenario depicted by Fernando Torres-Gil, gerontologist, age advocate, and former US Assistant Secretary of Aging, of millions of falling older bodies along with his call for action around fall prevention may already be with us. The gerontological literature portrays a frightening statistical picture of the costs, frequencies, injuries, hospitalizations, and deaths caused by falls among aging populations. In the United States, between 45 and 61 percent of nursing home residents fall each year, resulting in significant cases of bone fractures, with the mortality rates in the year following hip fractures alone estimated at between 14 and 35 percent.[27] Even for those who survive a hip fracture, many fail to recover normal bodily functioning. In Canada, falls are a major cause of injury, a reason for entering care facilities, a factor in morbidity, and an annual direct healthcare cost of $2.4 billion. According to Ward-Griffin et al., "Falls are the second leading cause of hospitalization in Canada for women 65 years and older and the fifth leading cause of hospitalization for men of this age."[28] Falls also cause 90 percent of all hip fractures for Canadian seniors, 20 percent of whom die within a year of the fracture.[29] Caring for older people injured in falls represents $1 billion of these costs.[30] In the United Kingdom, 50 percent of older residents in hospitals and care homes fall at least once a year.[31] While fall rates for community dwelling people 65 years and older is 30 percent, for those 80 years and older the rate is 40 percent, and falls are the most common cause of accidental death for those 75 years and older, with the total fall costs to the British National Health Service in the billions.[32] Fall rates vary by country and region; however, the World Health Organization's *WHO Global Report on Falls Prevention in Older Age* published in 2007 (hereafter referred to as the *WHO Report*) estimates that 28 to 35 percent of people 65 years and older fall each year, with the fall rate increasing with age.[33]

With so many older people falling, hospitalized, disabled, and even dying from their injuries, the risks of falling are obviously a formidable daily presence in the lives of older people and redraw the material boundaries between their quality of life and dependency, mobility, residence, and social support. These changes also affect the person's somatic status within cultural orders of health and well-being. Many studies show that the fear of falling creates tremendous anxiety for older people who can develop a resistance to physical activities because of it.[34] Thus, falls have profound consequences for biographical aging and a person's negotiation of identity in old age. However, in the public imagination, bodies that fall down also fall out of the social domain. They are assumed to belong to people who lack the physical control to "hold on" and thus become "fallers" out of balance with their environments. While falls are attributed to the older faller, as if their proclivity to falling was already a certainty, it is the fall that defines the faller. And this happens because a fall is an entry point into professional worlds of care, risk and prevention programs, hospital and community centers, and insurance and medical planning. Thus, the process of falling out of one world and into another is not simply physical; it is also embedded in the discursive practices and authoritative vocabularies that define the relationship between falls and fallers.

But what is a fall? While metaphorically, a fall connotes undeniable connotations of failing and loss,[35] the search for precise empirical definitions is fraught with inconsistencies. The most commonly reiterated definition is found in the *WHO Report*, where a fall means "inadvertently coming to rest on the ground, floor or other lower level, excluding intentional change in position to rest in furniture, wall or other objects." Furthermore, "many studies fail to specify an operational definition, leaving room for interpretation to study participants. This results in many different definitions of falls. For example, older people tend to describe a fall as a loss of balance, whereas health care professionals generally refer to events leading to injuries and ill health."[36] Why should "many different definitions of falls" be untenable, including self-reports by older people whose definitions include "loss of balance"? Why drive out "room for interpretation" for the sake of operational precision? And for whom are precise definitions of falls necessary?

Zecevic et al. compared the meanings of falls articulated in the research literature, particularly in the report by the Kellogg International Work Group on the *Prevention of Falls by the Elderly* (1987), with the views of older individuals and health workers. Since the report recorded falls as a disease,[37] with little consultation with older people about their embodied perceptions of falling, the authors conclude that "seniors talked more about the antecedents to falling and fall consequences, whereas researchers focused mainly on the description of the event itself and on exclusion criteria."[38] Loss of balance appears to be a major reason cited by people for falling because this is what it *feels* like to fall in an embodied sense in relationship to the external environment. However, the research community often treats loss of balance, along with "trips" or "slips," as less significant compared to the measurement of fall-related injuries.

Therefore, it is not surprising, as several studies indicate that risk-prediction programs and tools do not work because their assessment exercises are narrowly geared to the measurement of risk factors to the exclusion of personal responses to falling.[39] For example, Oliver reports in a study of fall-assessment tools in British hospitals that they provide only "one-off" measurements that offer individuals no long-term guidance about future prevention.[40] Oliver concludes that such tools are largely useless and recommends that if care providers "look after all older people in hospital better, it is likely they will fall less."[41] If research demonstrates an incompatibility between the inner, embodied narration of falling and its outer instrumental assessments, this should compel us to rethink the phenomenological nature of the relationship itself between fall and faller and take account of what a fall means to an older person.[42] This problem of incompatibility is evident in the larger field of fall prevention, where the materiality of embodied aging extends to political and gendered contexts as well.

FALL PREVENTION AND THE RISKS OF AGING BODIES

Given the magnitude and costs of falls, fall prevention for older people has grown into a sizable field of its own. It ties together the structural priorities of health governance, community living, and professional care with everyday concerns about lifestyle, behavior, environment, and intervention. Indeed, the literature on fall prevention characterizes the causes, treatments, and consequences of falls as "multifactorial" and typically lists the following as combined risk factors: medications (too many medications, adverse side-effects, improper mixing or administration of medications), environments (icy sidewalks, insecure rugs, loose flooring, poorly lit stairways), problems with assistive technologies (lack of or defective grab and transfer bars in bedrooms and bathrooms, handrails on stairways, working alarms), physiological problems (osteoporosis, declining muscle strength, poor vision, unstable posture), and psychological states (fear of falling, lack of confidence, poor self-care skills). Therefore, fall prevention programs advocate behavioral change (exercise regimes, healthy living, medication regulation) and environmental modifications (better lighting, nonslip surfaces, safe outdoor areas). These sets of multifactorial causes and preventative measures tend to dominate the professional organizations, such as the *Prevention of Falls Network Europe* (www.PROFANE.eu.org) or *The Falls Prevention Center of Excellence* in California (www.stopfalls.org/). The exercise and psychological factors are also core features of the many popular advice books currently on the market, with titles such as *Fall Prevention: Stay on Your Own Two Feet* (2006), *Fall Prevention: Don't Let Your House Kick You Out* (2006), and *How to Prevent Falls: Better Balance, Independence and Energy in 6 Simple Steps* (2007).

Although fall-prevention practitioners are devoted to improving the lives of older people, they also work within the discursive and practical limitations of their field. As a discourse, fall prevention works on balancing the collective statistical inventory of risks of aging against which individual lifestyles and health choices are seen to mitigate. Knowing that one-third of all people sixty-five years and older have fallen at least once each year is a collective risk based on age, a "fact" to be mediated by an individual's management of their bodies, homes, and medications. This double register of risk—collective and individual—is akin to the risk profiles described by Elizabeth Wheatley in her analysis of cardiac rehabilitation patients.[43] Wheatley calls the slippage from collective to individual risk a "truth trick" because the objective of rehabilitation is to foster individual control of cardiac health through changes in behavior and lifestyle. Thus "heart-smart" moral choices between "good" and "bad" rehabilitative behaviors override genetic, external, and financial determinants of health. Similarly, fall-prevention programs can be seen to individualize the risks of falling by emphasizing individual autonomy over structural factors such as income disparities and diminishing healthcare provisions. The *WHO Report* (2007) is clear that "active ageing" is the key strategy for fall prevention, but this strategy depends on adequate resources, supportive environments, and governmental commitment. Otherwise, fall prevention becomes another agenda of neoliberal health promotion campaigns and the coercive moralization of individual choice, responsibility, and risk-aversive lifestyles.

There are also myriad practical problems in the fall prevention field. For example, fall evaluations can be very complicated and involve the coordination of several professionals and providers who are not necessarily familiar with multifactorial conditions.[44] In addition, the requirements that some programs assess a person's home environment as well as the person's physical condition, combined with the confusing terms of Medicare coverage, can hamper the implementation of fall prevention. Assistive devices are another technical area where preventative solutions for falls, such as grab-bars, non-slip surfaces, easy-to-reach faucets and towel racks, evoke larger issues of social support, independent living, and affordable "age-friendly" housing. In Canada, despite the fact that 10–15 percent of all nonsyncopal (not due to sudden loss of consciousness) at-home falls happen in bathrooms, fewer bathtub grab-bars are installed in privately owned buildings than in buildings publicly owned. Since one-third of all Canadian seniors live in apartment buildings, this is a political issue and not simply one of individual choice to upgrade unsafe home environments.[45] The risk of bathroom falls is also affected by access to affordable high-quality bathtub transfer grab-bars, which range in quality, style, and expense. Here the materiality of embodied aging is contextualized within a microcosmic world of the grab-bar, the bathtub, the bathroom, the apartment, and the resident. These coalesce into an intimate assemblage of spaces and relationships through which an older individual's anxieties about assistance and dependence intersect with stratified levels of social support and care provisions.

Finally, gender is a crucial variable in fall prevention because the common image of women's bodies is that they are weaker, more vulnerable, and more risk-prone (especially after menopause due to osteoporosis) compared to those of men. If women do fall more often and suffer more fracture-related falls than do men, this is also a social consequence of the embodiment of asymmetrical gender relations and the cultural bias that depicts female physical strength as unfeminine and older female frailty as natural.[46] Women are often given a higher number of medications, and some falls may be attributable to women's use of psychotropic drugs and not their lack of physical endurance.[47] How men and women are differently expected to engage in physical risk further contributes to the overall bias. A British study of older individuals who have had two or more falls, notes that while more men perceive the risks of falling to be a matter of personal control, more women blame their own carelessness.[48] Further, as Horton and Arber illustrate in their survey of carers in fall-prevention behavior, female carers of older individuals tend to encourage autonomy, self-esteem, and reasonable risk-taking, while male carers practice a more protective and managerial caring style, especially where the care-recipient is female. Thus, gendered styles of risk-preventive caring not only affect the care-recipients' negotiation of identity between active autonomy and passive management, but also reproduce the social imbalance of power between men and women.[49] These factors, in turn, can intersect with generational differences in family authority.[50]

The discursive, technical, social, and gendered aspects of fall prevention together represent a larger configuration of aging bodies caught between the politics of risk and the governance of self-care. What older people think of participating in fall prevention programs is a good indicator of their own perceptions about this configuration. Many programs report low participation rates despite the programs' promotion of nonstigmatizing, "proactive," gender-sensitive, and socially conscious frameworks. When consulted about fall prevention programs, however, older people have been found to think of falls as less important than other health factors, and that the professional documentation is out of keeping with their own experiences of embodied aging.[51] Sometimes it is the unappealing advice to older women that they wear unattractive hip padding or special shoes, which disregards the women's own sense of embodied identity.[52] Many are also right to feel that fall prevention classes only increase their anxiety about falling and medicalize their lives.[53] Whether they are anxious about falls or not, for people who are judged to be at risk of falling, participation in a fall-prevention program means adding oneself to an at-risk-category that signifies old age and loss of control.[54] "The need to attend falling prevention classes, make safety modifications in the home, wear protective clothing or use mobility aids, may be viewed as announcing the transition to old age, dependence and further decline."[55] As Paul Kingston further notes, falling is "a powerful metaphor of decline" and a "status passage" because falls are embarrassingly publicized and become part of one's bio-identity.[56] If falling embodies aging in ways that signify decline, then it is not surprising that aging individuals attribute falls to accidents, environmental hazards, and the contingencies of life, rather than to the failing or frailty of their bodies.

Fall prevention discourses also frame any risk-taking, which is assumed to be a natural part of younger lives, as being dangerous and foolish in later life. In the community, some risk-taking may be beneficial and an important aspect of negotiating boundaries around activity, competence, autonomy, and social interaction. For example, in the case of an older person who lives alone, the social routine of visiting neighbors or shopping on a wintry day to prevent housebound isolation should be weighed against the potential risks of walking on snowy sidewalks. In care programs, the exclusive attention to reducing physical risks may actually cut out benign or healthful risk-taking behaviors. In this regard, Ballinger and Payne's research on a day facility in Britain finds that "the focus on physical activity and risk, and the staff's role in limiting 'inappropriate' patient behaviour, reinforced passivity among the service users."[57] At the same time, older service users were concerned with social rather than physical risks, such as "being ostracised through unexpected or unconventional behaviour, the risk of infantalisation on the part of health professionals, and the risk of being stigmatised in one's local community."[58] If our society expects its older citizens to be responsibly self-caring about their health, then we also have to realize that in addition to the wisdom that comes with cautious adjustment to changing physical and material conditions of aging, are those life skills sustained through reasonable experimentation, taking chances, and exploring new realms of somatic experience.

CONCLUSION: PHYSICAL AND CULTURAL AGING

This chapter's discussion on falls serves the wider goal of locating aging bodies in the nexus between physical, biographical, and cultural realms, where embodiment becomes the material ground from which to challenge both biological and constructivist reductionism. In so doing, the chapter contributes four theoretical points to the conversation between aging studies and body studies and to the interdisciplinary task of bringing aging bodies into critical focus.

First, falls are indeed "multifactorial," but more importantly they are multidimensional because they point to a horizon of relationships connecting aging bodies to vital social worlds of health, safety, housing, and quality of life. The single moment of a "hard" fall shakes up this horizon and potentially dislocates a person from their environments, routines, and networks of activity. Indeed, while falls just happen, they create fallers and symbolize the stigmatizing passage into old age.

Second, against an ageist society obsessed with fears of decline and loss of physical control, older people contend with fall-related injuries by relying on their subjective, interpretive, and adaptive resources, as well as family, medical, and community supports. What falls mean is as important as what falls do, and older people's embodied perceptions about risks should be part of risk-prediction and fall prevention programs. This is especially significant because risk prevention and health promotion programs charge older individuals with being responsible for their own well-being and condemn risky activities as being either irresponsible or age-inappropriate behavior.

Third, falls provide a point of multidisciplinary convergence for knowledge-making about aging in the sciences, the professions, the social sciences, and the humanities. As each of these navigates the conceptual, practical, and ethical corridors that separate autonomy from dependency in old age, the magnitude of issues around falling aging bodies provides an exciting opportunity for these fields to find common ground. Falls cannot be completely defined, medicalized, operationalized, predicted, or prevented, yet their complexity tells us that treatment and intervention have to take into account social, gendered, technological, and political determinants of health in old age.

Fourth, falls do tell gerontological stories, to return to Mike Hepworth's observation, which push aging bodies to the forefront of our understanding of the human spirit. The older body that falls or is at-risk of falling is a portal

from which to view the contingent nature of the aging process, and the
ways in which biography, culture, politics, and biology are braided together.
Just as the embodiment of aging has a subjective dimension, the subjectivity of aging has a physical dimension that is materialized in the activities,
environments, and social systems in which we grow older. The dynamic
relationship between these dimensions is what constitutes the experience of
aging and contests the imposition of universalizing scientific and cultural
narratives. Falls trace a new folding in the life course, open an elastic fron-
tier between the inside and outside of aging, and narrate how vulnerability, chance, suffering, fear, imagination, and strength interact in so many
embodying ways.[59]

NOTES

1. Mike Hepworth, *Stories of Ageing* (Buckingham: Open University Press, 2000), 9.
2. Kathleen Woodward, *Aging and Its Discontents* (Bloomington and Indianapolis: Indiana University Press, 1991), 19.
3. The critique of Third Age retirement culture as a celebration of "empower-ment," consumerism, and leisure has been dealt with elsewhere. See Kevin E. McHugh, "'The Ageless Self'? Emplacement of Identities in Sun Belt Retirement Communities," *Journal of Aging Studies* 14 (2000): 103–115; Margaret Morganroth Gullette, *Aged By Culture* (Chicago: University of Chicago Press, 2004); Chris Gilleard and Paul Higgs, *Contexts of Ageing: Class, Cohort and Community* (Cambridge: Polity Press, 2005); Stephen Katz, *Cultures of Aging: Life Course, Lifestyle and Senior Worlds* (Peterborough, ON: Broadview Press, 2005).
4. Paul Higgs and Ian Rees Jones, *Medical Sociology and Old Age* (London and New York: Routledge, 2009), 95.
5. Peter Öberg, "The Absent Body—A Social Gerontological Paradox," *Ageing & Society* 16, 701–719; Peter Öberg, "Images versus Experience of the Aging Body," in *Aging Bodies: Images and Everyday Experience*, ed. Christopher A. Faircloth (Walnut Creek, CA: Altamira Press, 2003), 103–139.
6. Sarah Nettleton and Jonathan Watson, eds., *The Body in Everyday Life* (London and New York: Routledge, 1998); Helen Thomas and Jamilah Ahmed, eds. *Cultural Bodies: Ethnography and Theory* (Malden, MA: Blackwell, 2004); Nick Crossley, *Reflexions in the Flesh: The Body in Late Modern Society* (Milton Keynes: Open University Press, 2006); Alan Petersen, *The Body in Question: A Socio-Cultural Approach* (London and New York: Routledge, 2007); Chris Shilling, ed., *Embodying Sociology: Retrospect, Progress and Prospects* (Malden, MA: Blackwell, 2007); Bryan S. Turner, *The Body and Society: Explorations in Social Theory (3e)* (London: Sage, 2008).
7. For example, Christopher A. Faircloth, ed., *Aging Bodies: Images and Everyday Experience* (Walnut Creek, CA: Altamira Press, 2003).

8. Dana Rosenfeld, *The Changing of the Guard: Lesbian and Gay Elders, Identity, and Social Change* (Philadelphia: Temple University Press, 2003); Julie Jones and Steve Pugh, "Ageing Gay Men: Lessons from the Sociology of Embodiment," *Men and Masculinities* 7 (2005): 248–260; Kathleen F. Slevin, "The Embodied Experiences of Old Lesbians," in *Age Matters: Realigning Feminist Thinking*, eds. Toni M. Calasanti and Kathleen F. Slevin (London and New York: Routledge, 2006), 247–268.

9. Anne Davis Basting, *The Stages of Age: Performing Age in Contemporary American Culture* (Ann Arbor: University of Michigan Press, 1998); Liz Schwaiger, "Performing One's Age: Cultural Constructions of Aging and Embodiment in Western Theatrical Dancers," *Dance Research Journal* 37 (2005): 107–120; Kathleen Woodward, "Performing Age, Performing Gender," *NWSA Journal* 18 (2006): 162–189.

10. Thomas R. Cole, Robert R., Kastenbaum, and Ruth E. Ray, eds., *Handbook of the Humanities and Aging* 2 ed. (New York: Springer, 2000); Thomas R. Cole, Ruth E. Ray, and Robert R., Kastenbaum, eds., *A Guide to Humanistic Studies in Aging* (Baltimore: Johns Hopkins University Press, 2010).

11. Kathleen Woodward, ed., *Figuring Age: Women, Bodies, Generations* (Bloomington and Indianapolis: Indiana University Press, 1999); Margaret Cruikshank, *Learning to Be Old: Gender, Culture, and Aging* (New York: Rowman and Littlefield, 2003); Samantha Holland, *Alternative Femininities: Body, Age and Identity* (Oxford and New York: Berg, 2004); Toni M. Calasanti and Kathleen F. Slevin, eds. *Age Matters: Realigning Feminist Thinking* (London and New York: Routledge, 2006); *Current Sociology* (2007), Special issue on Changing Approaches to Gender and Aging, 55(2); Laura Hurd Clarke, *Facing Age: Women Growing Older in an Anti-Aging Culture* (New York: Rowman and Littlefield, 2010).

12. Barbara L. Marshall and Stephen Katz, "From Androgyny to Androgens: Re-Sexing the Aging Body," in *Age Matters: Realigning Feminist Thinking*, eds. Toni M. Calasanti and Kathleen F. Slevin (London and New York: Routledge, 2006), 75–97; Celia Roberts, *Messengers of Sex: Hormones, Biomedicine and Feminism* (Cambridge: Cambridge University Press, 2007).

13. Clary Krekula, "The Intersection of Age and Gender: Reworking Gender Theory and Social Gerontology," *Current Sociology* 55 (2006): 155–171.

14. Allison James and Jenny Hockey, *Embodying Health Identities* (Basingstoke: Palgrave Macmillan, 2007), 39.

15. Simon J. Williams, "Beyond Meaning, Discourse and the Empirical World: Critical Realist Reflections on Health," *Social Theory & Health* 1 (2003): 42–71; Myra J. Hird, "Re(pro)ducing Sexual Difference," *Parallax* 8 (2002): 94–107; Chris Shilling, *The Body in Culture, Technology and Society* (London: Sage, 2005).

16. Toni Calasanti, "Ageism, Gravity, and Gender: Experiences of Aging Bodies." *Generations* 29 (2005): 8–12; Julia Twigg, *Bathing—The Body and Community Care* (London and New York: Routledge, 2000); Julia Twigg,

The Body in Health and Social Care (Basingstoke: Palgrave Macmillan, 2006); Emmanuelle Tulle, *Ageing, the Body and Social Change* (Basingstoke: Palgrave Macmillan, 2008).

17. Twigg, *The Body in Health and Social Care*, 25.
18. Maurice Merleau-Ponty, *Phenomenology of Perception* (London: Routledge and Kegan-Paul, 1962); Pia Kontos, "Embodied Selfhood: An Ethnographic Exploration of Alzheimer's Disease," in *Thinking about Dementia: Culture, Loss, and the Anthropology of Senility*, eds. Annette Leibing and Lawrence Cohen (New Brunswick, NJ: Rutgers University Press), 2006), 195–217; Eric Matthews, "Dementia and the Identity of the Person," in *Dementia: Mind, Meaning and the Person*, eds. Julian C. Hughes, Stephen J. Louw, and Steven R. Sabat (New York: Oxford University Press, 2006), 163–178.
19. Twigg, *Bathing—The Body and Community Care*, 46.
20. Simon Biggs, *The Mature Imagination: Dynamics of Identity in Midlife and Beyond* (Buckingham: Open University Press, 1999).
21. Simon Biggs, "Negotiating Aging Identity: Surface, Depth, and Masquerade," in *The Need for Theory: Critical Approaches to Social Gerontology*, eds. Simon Biggs, Ariela Lowenstein, and Jon Hendricks (Amityville, NY: Baywood, 2003), 156.
22. Steven Wainright and Bryan S. Turner, "Aging and the Dancing Body," in *Aging Bodies: Images and Everyday Experience*, ed. Christopher A. Faircloth (Walnut Creek, CA: Altamira Press, 2003), 259–292; Steven Wainright and Bryan S. Turner, "'Just Crumbling to Bits'? An Exploration of the Body, Ageing, Injury and Career in Classical Ballet Dancers," *Sociology* 40 (2006): 237–255.
23. Other studies of dance illustrate the materiality of embodiment taking different forms and becoming the basis for new physical skills. See Schwaiger's study of body memory and "intercorporeity" among mature dancers; Liz Schwaiger, "The Flesh and the World: Intercorporeal Body-Selves, Ageing and Dancing," *Senses and Society* 3 (2008): 45–59.
24. Tulle, *Ageing, the Body and Social Change*; Emmanuelle Tulle, "The Ageing Body and the Ontology of Ageing: Athletic Competition in Later Life," *Body & Society* 14 (2008): 1–19.
25. Tulle, "The Ageing Body and the Ontology of Ageing," 15.
26. Fernando Torres-Gil quoted in Paul Kleyman, "Safety in the Balance: Preventing Falls and Injuries in Elders. Beyond Grab Bars: Adapting to Reality," *Aging Today* 29 (2008): 7, 10.
27. Norman V. Carroll, Jeffrey C. Delafuente, Fred M. Cox, and Siva Narayanan, "Fall-Related Hospitalization and Facility Costs among Residents of Institutions Providing Long-Term Care," *Gerontologist* 48 (2008): 213.
28. Catherine Ward-Griffin, Sandra Hobson, Pauline Melles, Marita Kloseck, Anthony Vandervoort, and Richard Crilly, "Falls and Fear of Falling among Community-Dwelling Seniors: The Dynamic Tension between Exercising Precaution and Striving for Independence," *Canadian Journal on Aging* 23 (2004): 308.

29. *The Council on Aging of Ottawa Bulletin*, "Special Home Safety Issue," (March 2006) http://www.coaottawa.ca/library/publications/bulletins/MarchBulletin06.pdf, 3. Accessed 11/23/2008.

30. Cathy Bennett, "Prevention Supports Seniors' Health and Independence," Senior Care Canada (2002) http://seniorcarecanada.com/articles/2002/q3/falls.prevention). Accessed 11/23/2008.

31. David Oliver, "Falls Risk-Prediction Tools for Hospital Inpatients: Time to Put Them to Bed?," *Age and Ageing* 37 (2008): 248.

32. The British Geriatrics Society, "Falls" (2007) http://www.bgs.org.uk/Publications/Compendium/compend_4-5.htm. Accessed 11/23/2008.

33. *WHO Global Report on Falls Prevention in Older Age* (Geneva Switzerland, World Health Organization, 2007).

34. See, for example, Nancy Edwards, Donna Lockett, Faranak Aminzadeh, and Rama C. Nair, "Predictors of Bath Grab-Bar Use among Community-Living Older Adults," *Canadian Journal on Aging* 22 (2003): 217–227; Nancy Edwards, Nancy Birkett, Rama Nair, Maureen Murphy, Ginette Roberge, and Donna Lockett, "Access to Bathtub Grab Bars: Evidence of a Policy Gap," *Canadian Journal on Aging* 25 (2006): 295–304.

35. Sandy O'Brien Cousins, and Donna Goodwin, "Balance Your Life! The Metaphors of Falling," *WellSpring* 13 (2002): 6–7.

36. *The WHO Report*, 1.

37. Aleksandra A. Zecevic, Alan W. Salmoni, Mark Speechley, and Anthony A. Vandervoort, "Defining a Fall and Reasons for Falling: Comparisons among the Views of Seniors, Health Care Providers, and the Research Literature," *Gerontologist* 43 (2006): 368.

38. Zecevic et al., "Defining a Fall and Reasons for Falling," 371.

39. Anne H. Laybourne, Simon Biggs, and Martin C. Finbarr, "Falls Exercise Interventions and Reduced Falls Rate: Always in the Patient's Interest?," *Age and Ageing* 37 (2008):10–13.

40. Oliver, "Falls Risk-Prediction Tools."

41. Ibid., 250.

42. For an example of research that incorporates the meaningful stories of people who have fallen, see Marianne Mahler and Anneli Savnimäki, "Indispensible Chairs and Comforting Cushions—Falls and the Meaning of Falls in Six Older Persons' Lives," *Journal of Aging Studies* 24 (2010): 88–95.

43. Elizabeth E. Wheatley, "Disciplining Bodies at Risk: Cardiac Rehabilitation and the Medicalization of Fitness," *Journal of Sport and Social Issues* 29 (2005): 198–221.

44. Mary E. Tinetti, Catherine Gordon, Ellen Sogolow, Pauline Lapin, and Elizabeth Bradley, "Fall-Risk Evaluation and Management: Challenges to Adopting Geriatric Care Practices," *Gerontologist* 46 (2006): 717–725.

45. Edwards et al., "Access to Bathtub Grab Bars."

46. Amanda Grenier, "The Distinction between Being and Feeling Frail: Exploring Emotional Experiences in Health and Social Care," *Journal of Social Work Practice* 20 (2006): 299–313; Amanda Grenier and Jill Hanley,

"Older Women and 'Frailty': Aged, Gendered and Embodied Resistance," *Current Sociology* 55 (2007): 211–228. An American study reports that older women suffer fall-related injuries at rates 40–60 percent higher than men of similar age and experience 81 percent higher hospital admission rates compared to older men. Nancy Nachreiner, Mary J. Findorff, Jean F. Wyman, and Teresa McCarthy, "Circumstances and Consequences of Falls in Community-Dwelling Older Women," *Journal of Women's Health* 16 (2007): 1437–1446.

47. Outside of residents in long-term care facilities in North America, the top category of drug users are women over the age of sixty-five, 12 percent of whom take ten or more medications. Judy Steed, "Drugged-Out Seniors a Prescription for Disaster," *Toronto Star* (November 11, 2008): A1, A13.

48. Khim Horton, "Gender and the Risk of Falling: A Sociological Approach," *Journal of Advanced Nursing* 57 (2006): 69–76.

49. Khim Horton and Sara Arber, "Gender and the Negotiation between Older People and Their Carers in the Prevention of Falls," *Ageing & Society* 24 (2004): 75–94.

50. See Christine Kilian, Alan Salmoni, Catherine Ward-Griffin, and Marita Kloseck, "Perceiving Falls within a Family Context: A Focused Ethnographic Approach," *Canadian Journal on Aging* 27 (2008): 331–345.

51. Barbara J. Braun, "Knowledge and Perception of Fall-Related Risk Factors and Fall-Reduction Techniques among Community-Dwelling Elderly Individuals," *Physical Therapy* 78 (1998): 1262–1276; Claire Ballinger and Sheila Payne, "The Construction of the Risk of Falling among and by Older People," *Ageing & Society* 22 (2006): 305–324; Frances Bunn, Angela Dickinson, Elaine Barnett-Page, Elizabeth McInnes, and Khim Horton, "A Systematic Review of Older People's Perceptions of Facilitators and Barriers to Participation in Falls-Prevention Interventions," *Ageing & Society* 28 (2008): 449–472.

52. Lucy Yardley, Margaret Donovan-Hall, Katherine Francis, and Chris Todd, "Older People's Views of Advice about Falls Prevention: A Qualitative Study," *Health Education Research* 21 (2006): 512.

53. Fall anxiety can sometimes be assuaged through nonmedical means, such as religious participation. See Carlos A. Reyez-Ortiz, Hanah Ayele, Thomas Mulligan, David V. Espino, Ivonne M. Berges, and Kyriakos S. Markides, "Higher Church Attendance Predicts Lower Fear of Falling in Older Mexican-Americans," *Aging & Mental Health* 10 (2006): 13–18.

54. Physiotherapy or similar therapies that focus on posture, gait, and muscle-strengthening exercises without reference to age and in mixed-age environments have been found to be less stigmatizing and more effective in terms of risk prevention. Unfortunately, physiotherapy coverage is not always covered by medical insurance plans. How allied health professionals such as physiotherapists and occupational therapists can better work together in fall prevention programs is an important issue of its own. See Samuel R. Nyman and Claire Ballinger, "A Review to Explore How Allied Health Professionals

Can Improve Uptake of and Adherence to Falls Preventions Interventions," *British Journal of Occupational Therapy* 71 (2008): 141–145.

55. Yardley et al., "Older People's Views of Advice about Falls Prevention," 509.

56. Paul Kingston, "Falls in Late Life: Status Passage and Preferred Identities as a New Orientation," *Health* 4 (2000): 218.

57. Ballinger and Payne, "The Construction of the Risk of Falling," 319.

58. Ibid., 320.

59. I wrote this chapter in the months following a fall story of my own. While away at a cottage in the French countryside, I fell one night from a small crest into a ditch behind it. It was dark and I was momentarily alone. I subsequently discovered that my knee was badly injured and I had to undergo physiotherapeutic care for several months. However, the fearful experience of being put to ground and "down," neither able to "hold on" nor "get up," was as penetrating as the pain in my knee.

CHAPTER 13

THE GIMMICK: OR, THE PRODUCTIVE LABOR OF NONLIVING BODIES

GEORGE SANDERS

> The expression "dead labor," by which one denotes capital, does not erase the fact that all labor is living labor.[1]

COUNTER TO BRUNO GULLI'S CLAIMS, above, the dead do in fact continue to labor. There are always *active* remainders; traces of a life once lived: "The social corpse is imbued with presence and personhood."[2] For some, the dead body provides a kind of serene comfort and reassurance that, even if not all is right in the world, there is at least some verisimilitude of a comforting presence. For others, the dead body is a source of terror. Either way, the lifeless body labors—it produces something—even if that something is only an affect or an affectation for abjection. Bodies continue after death to produce attachments, aversions, and other emotions.

This labor cannot properly be considered to be *living* labor, but is it what Marx called "dead labor"? And can it find fungible purchase as such? Can the dead body be subsumed into capital's circulatory system? In short, yes, and in this chapter I describe the transformation of the laboring dead body into the *gimmick*. The gimmick is a mechanism contrived to draw the living into monetized relations. It is an event-structure that relies on surplus value (labor) even as it simultaneously converts that labor into an exchangeable commodity (product). In the funeral trade, the gimmick exploits the labor of the dead and transforms the dead into something that can be bought and sold. Capital inscribes (dead) bodies to create new relationships—relationships between the living and the dead, to be sure, but also reflexive relationships within the selves of the living—with one's experiences of time, memory, and emotion.

The body is, among other things, symbolic of time. Just as bodies inevitably grow and vanish, so do relationships (if only as a consequence of human mortality). In death studies, the concept of liminality is usually invoked to discuss the period of time following a death in which bereaved persons resituate their social identities (e.g., a wife to widow). Thus, the body (what it does and what we do with it) is an organic timekeeper that helps guide human relations. Liminality, though, represents a forestalling of time, a "no-man's land betwixt and between,"[3] during which time identities and relationships become disconfigured and open-ended.

The gimmick has an elective affinity for liminality since a gimmick is designed to create continuous but unfulfillable desire, typically desire for a particular commodity. So, just like liminality, the gimmick engenders neither the complete absence of a desire nor the satisfaction of it—and, like liminality, it is temporally bounded (at least, conventionally). Importantly, some commodities can themselves be gimmicks, since they promise nothing more than their own empty (yet potent) promises. Some commodities do little more than create the desire for something else, endlessly deferring any consumptive gratification one might expect to have received.

The funerary gimmick is just such a type. The transformation of the dead body into a gimmick exploits the labor of the dead, even as it creates the desire for itself without ever allowing the consumer to realize the consummation of that desire. The false promise entailed by desire in unison with the ephemeral and parodic nature of the gimmick transforms the dead body into a source of surplus value within the residue of a life once lived.

PROMISES, PROMISES

Until recently, dead bodies were, for the most part, discarded—either buried or burnt and pulverized into ashes that were then compartmentalized in niches, scattered, or, in a minority of cases, at least in the United States, stored on mantles. Today, of course, tissues can be excised and organs reintroduced into living bodies. And there are an increasing number of novel technologies that allow a growing pool of entrepreneurs to process dead bodies in previously unheard of ways. More than one-third of all bodies in the United States are cremated, and the rate is expected to rise above one-half in another generation.[4] As a result, so-called end-trepreneurs[5] have stormed onto the funeral industry scene, making available everything from themed funeral services and ash-scattering parties to branded urns and products made out of cremated remains, also known as "cremains."

What many of these goods and services have in common is their *spectacular* nature. Oliver McRae, who writes often for the premier publication of the National Funeral Director's Association, *The Director*, on economic issues contends, "The living have no interest in buying a burial, but they are very interested in a glorious celebration of the beauty, magic, wonder and spectacle of life, with friends and family and love all around, i.e., a powerful and poetic funeral."[6] McRae's use of the term *spectacle* is fortuitous since it might (albeit unintentionally) be associated with Guy Debord's claim that we in the West have come to inhabit a society of the spectacle where the citizenry has fallen under the spell of sensationalism, constant imagistic flourishes, and an overall media smorgasbord filled with visual hyperbole.

Debord writes, "[T]he spectacle is both the meaning and the agenda of our particular socio-economic formation... [It] is the present model of socially dominant life."[7] He argues that the expansion of capital is made possible through the creation of sophisticated imagery that appealed to citizens on a visceral, rather than cognitive level. This would suggest that people do not consume goods for their content but primarily for their presentation. The endless procession of empty promises accompanying the pageantry of the spectacle invariably carries consequences. Baudrillard argues that the spectacle assumes its own reality.[8] It is more real than any other reality because we treat simulations, our spectacles, as though they do more than represent. That is, the spectacle is the *really* real—it has become the reference point by which other dimensions of reality are assessed. The signifier has become the referent. And consumers, Baudrillard claims, now accept promises as just that—mere promises, or empty locutions. This brings us to the gimmick, which itself is a kind of empty promise.

When accepted as vacuous assurance, the gimmick is entirely liminal. It presumes a subject who is incomplete and for whom completion is *impossible*. From its etymology we learn that *liminal* is related to the *limen*, the threshold, but perhaps one might as easily relate it to *limn*—to trace or outline. The threshold, just like the outline or trace, exists in the order of the (cultural) imaginary. Neither inside nor outside, it is an empty promise, fixed permanently betwixt-and-between points, a lá Zeno's paradox. This liminal space that simultaneously contains and frustrates desire revivifies the need for some solid ground and comforting certainty—something the body seems to provide.[9] Yet even the body (whether living or dead) is not impervious to the betwixt-and-between of capital's gimmickry.

DISMEMBERING BODIES

The gimmick is a particular manifestation of the twenty-first-century spectacle. Like the spectacle, the gimmick's reality is diffuse, imaginary, and often self-referential and insulating. The gimmick is a unique expression of late capitalism since it is designed to appeal to both authenticity and irony simultaneously. This is evident throughout today's funeral industry in the United States but especially when it comes to the novel processing of dead bodies.

LifeGem, founded in 2001, is probably the most well known of the companies that manufacture funerary gimmicks. The suburban Chicago-based company processes the cremated ashes of a dead body to make wearable jewelry. Human bodies are, after all, essentially carbon-based life forms and, when exposed to extreme, prolonged heat the carbon becomes highly concentrated in the resulting ash. Diamonds, too, are carbon artifacts that have undergone heat and pressure. Enter LifeGem, whose founders invented a process whereby human ashes can be turned into synthetic diamonds.

So many people are aware of LifeGem (as opposed to some of the other companies I discuss below) because of the company's promotional campaigns. Since all gimmicks need a public, and given the fact that it was one of the original companies to transform the dead body into a gimmick within the funeral industry, LifeGem has garnered much attention from major media outlets. One need merely click on the company's webpage, on the section "In the news…" (http://www.lifegem.com/secondary/LifeGemMedia2006. aspx) to sample the wide-ranging media coverage about the company. There, the company links to various news stories and video clips of people discussing LifeGem's products.

Of note, these web pages are more typically used to highlight the glowing comments of putatively impartial reviewers who lavish praise on the company's product-line. But on LifeGem's "In the news…" page, the majority of stories include satirical and even sardonic remarks. Indeed, it is sometimes difficult to tell whether the reporter is denigrating the company and its products or advocating for their universal adoption.

The fact that LifeGem links to so many media stories that parody its product suggests the company is not unaware of the gimmickry of its body-as-commodity/labor. In other words, the company uses parody as a sales tactic, albeit one that intrinsically *hedges* its promises to the consumer. Therein lies the gimmick—by implying that the buyer may not end up *liking* the product, she will, at the very least, discover some pleasure in the absurdity of having delivered the dead body of a loved one over to a shill. This leverages liminality by exposing the otherwise concealed quality of the consumable as existing in a region that neither offers complete satisfaction nor depletes the consumer's desires for more and more goods.

It is worthwhile to witness the gimmick at work in another way. Rather than utilizing public commentary as the gimmick to create desire for the commodity, the dead body itself has become the gimmick, which is used to metonymize the company. Besides a link that opens up an excerpt from a pre-dictable one-liner during a Jay Leno monologue, there is also a link to news commentator Keith Olbermann's remarks. Olbermann acerbically says,

> It was, we all thought, bad enough when they froze baseball immortal Ted Williams or when a Houston company launched the last mortal remains of Dr. Timothy Leary into geosynchronous orbit and offered to do the same for you but only when you were dead...We all thought those were bad enough but as "Countdown's" Monica Novotny reports...we were all wrong!

There's even a link to a bit from "Live with Regis and Kelly" in which Kelly Ripa says, "It seems a little creepy to me" and proceeds to joke that in order to get a tennis bracelet (which requires many diamonds and, thus, many husbands) she would have to become a "black widow." Out on the "Today Show" plaza, one newsperson states, "It's something you can do with your loved ones after they've passed away," to which the jokester/weatherman, Al Roker, replies, "Wow, you're a real jewel." Even the venerable Scott Simon of NPR jokes with one of the company's founders that it is now possible for "diamonds really to be a girl's best friend."

On the other hand, a few remarks seem to support a more charitable interpretation of the company and its products. One customer interviewed on *Inside Edition* is asked to speculate on how the deceased might feel about her cremains being processed into wearable jewelry. He responds, "She'd think this is terrific. She's looking down smiling." His reply feels utterly uncontrived and seems to represent an unqualified desire to authentically reconnect with his dearly departed. A few other media accounts depict customers who justify their purchases because they want to be "close" with their (deceased) loved ones. More often than not, though, some degree of irony is present in the views expressed in the outlets linked to the company's web page. All of which attests to the subservient role of the dead body. Where once it marked time for funeral participants, now it eschews time for consumers.

MODULATING MORTALS

Is the dead body (as gimmick) simply one more area of "life" that is subsumed by the circulation and aggregation of capital such that "Older capacities of the human body are reinvented, new capacities revealed"?[10] Yes and no. The gimmick differs from previous versions of the commodity form since it so readily leverages the liminal. The gimmick is not merely an additive dimension of commercialization of the body but a qualitatively distinct kind of commercialization. For as long as there has been a funeral industry, there has been some degree of commercialization with respect to funerary goods and services. However, the gimmick represents not merely the commodification of a ritual and a dead body; it also represents the commodification of a temporal nether-region that brackets these things. That is, the funerary gimmick puts the waxing and waning of the ritual and the disposal of the body in permanent holding patterns. The gimmick places grief in abeyance. Where "the grieving self *is* the 'negative self of desire,'"[11] the consuming self is free to be the site for the production of ever-new (and ever-unsatisfiable) desires.

Historically, funerary ritual (with all of its corresponding goods and services) is meant to mark the impermanence of liminality by explicitly highlighting the impermanence of the body. It is only during the liminal (mourning) period when the desires of the bereaved are suspended between two opposing attractors. On the one hand they, the bereaved, wish to let go and leave be. On the other hand, they wish to retrieve that which has been lost. The bereaved person is stuck in a sort of limbo both wanting to reunite with the dead and to completely sever all ties.[12]

However, funerary ritual demonstrates that as the body moves from a state of alive-ness to morbidity, bereaved persons are likewise expected to move from one state to another. The casket is shut (or the crematorium fire is lit). A handful of dirt is tossed into a grave (or ashes are sealed in an urn). End of story. There is a temporal progression that takes place in which the body of the deceased symbolizes the social body. Bereaved persons are expected to mourn their loss and move on with their lives (albeit lives with new social configurations). The gimmick, however, is itself a liminal. Rather than facilitating the continued circulation of relations, the commodity gimmick freezes the cyclical nature of biologically necessitated rituals.

The fact that the dead body can be a site for engaging liminality (with its suspended desires) is evidenced by the bereaved who express the need to see and commune with the dead body. Consider the entreaties by mourners following the attacks on September 11, 2001, for the recovery and identification of those who had been killed; or the trauma endured over years by the countless survivors of *desaparecidos*. In an interview-based study with bereaved persons, Gentry et al. quoted individuals as saying things such as "I liked holding her hand for three hours—friends came and went...I spent much time at the funeral home, and stayed at the gravesite until the body was lowered. I feel that I acknowledged the finality of her death; I wanted to make sure that I never had any phantom experiences where I was expecting her to come around the corner."[13] The time spent with the body eventually comes to an end. The visage of the deceased is gazed upon, the body is touched or held, and then the body is buried or cremated. The funerary gimmick, however, has its own shelf-life, one that may not necessarily conform to the cyclical nature of grief and remembrance. Instead, it creates its own history.

HUMORLESS HUMOR

In *Looking Awry*, Zizek says that the "fundamental fantasy of contemporary mass culture" is the "return of the living dead."[14] The funerary gimmick appeals to this "fantasy" by making available multiple outlets through which the dead may, as it were, "return," albeit in the form of labor subject to the machinations of capital. The success of LifeGem has spawned a number of direct competitors, including GemSmart and New Life Diamonds in the United States. There are also a number of companies around the globe offering wearable jewelry made from processed cremains, suggesting transnational elements to these socioeconomic dynamics of death.

Probably second to LifeGem in terms of popular media coverage is Eternal Reefs, a company that uses human ashes as a base material for artificial reefs. Cremains are mixed with other substances to create a sphere between two and six feet in diameter. The funeral service consists of a boat trip to a location at which the reef is lowered onto the sea floor and loved ones are provided with GPS coordinates for future snorkeling visits. The company links to quite a number of news reports (http://www.eternalreefs.com/about/news.html) discussing the offerings. Most of these reports focus on the environmental-friendliness of the company and its products, but there are still plenty of puns (e.g., *Wired*'s story is called "Deep-Sixing your Ashes" and *Business Week*'s article is entitled "Sleeping with the Fishes—Forever"). Not surprisingly, Eternal Reefs has a major competitor—Neptune Memorial Reef.

Both Eternal Reefs and Neptune Memorial Reef provide the consumer not only with a material product (using the dead body as its base), they also provide the consumer with prepackaged rituals (the bereaved sail off shore to a drop point and hold a funeral service on arrival). While these rituals do provide a temporal structure, some might see their ability to move the bereaved into and out of a liminal period as superficial. One sociologist writes, "American funerals are not very sad occasions ... the funeral has become an empty, shallow, and increasingly worthless ritual,"[15] and some psychologists, too, weigh in: "The enactment of funeral and bereavement rituals in contemporary American culture is often inauthentic, a hollow and rigid practice, devoid of an opportunity for genuine healing."[16] The use of gimmicks in funerary rites is often justified by a desire to escape turgid, torpid, and arbitrary rituals in which one merely goes through the motions. However, others feel that gimmicks perpetuate rather than diminish any perceived emptiness.

Some companies offer only the dead body as a material good, with varying levels of utility. Most of the products I discuss in this section preserve the body in a sort of mummified state. The body's status as commodity, though, problematizes the labor that an unpreserved body accomplishes. One "owns" a dead body whose "ontological status constitutes a rupture in the very notion of progression from 'here' to 'there.'"[17] However, with these novel products there is only a "here."

Companion Star Crystal, for instance, incorporates cremains into hand-blown crystal vases, jewelry, or pendants, and artist Beth Menczer mixes cremains with clay and creates sculptures. Spone Funerary Ware combines cremains with porcelain to create "human bone china" in the form of plates or vases. Of all of the companies I studied, Relict Memorials is likely the most implicitly poststructural. The company uses cremains to create a memorial headstone of the deceased, transforming the dead body into pure cipher. But it's not alone in its penchant for semiotic self-referentiality. Honor Industries mixes cremains with other materials to create charcoal pencils. An artist is then hired to sketch a portrait of the deceased with the pencils. Similarly, Ashes to Portraits combines cremains with oil paints and hires someone to paint a picture of the deceased with the mixture. The company now has a competitor with the birth of similarly (and confusingly) named Portraits from Ashes.

Each of these companies transforms the dead body into something wholly different—something consumable, yes, but something also that lacks the unpleasant properties typically associated with the dead body (decay, rot, ooze). As Zizek writes, "On today's market, we find a whole series of products deprived of their malignant properties: coffee without caffeine...the Colin Powell doctrine of warfare with no casualties...tolerant liberal multicultur-alism as an experience of the Other deprived of its Otherness."[18] Indeed, it is the property of abjection and all that comes with it that comprises such a malignant property, but it is also the "malignant" property of waste.

For capitalism, waste or "surplus" is a problem that must be redressed. Given the axiom that profit is sought for the sake of profit, there is, conse-quentially, an ever-present, looming threat of excess that must be somehow reabsorbed. Otherwise, the mode of production risks making apparent one of capitalism's many legendary contradictions. Where surplus capital trans-lates into surplus labor there are surplus bodies, and bodies are consigned to instruments of exchange. Liminality is, by definition, a form of surplus. It refuses to fit into a particular identity, temporal era, or truth-claim. Its "betwixt-and-between" epitomizes a neither-nor. Liminality around death represents a bracketing of time that exists outside of the boundaries of the workaday world. Someone who is paralyzed with grief can neither be a pro-ductive worker nor a virtuous consumer.

Late capitalism demands the ever-onward march of commodification, leaving no bodies untouched. This includes dead bodies, which are only made to *appear* to lack productive labor. Living bodies have become incapable of producing their own selves/subjects without the subjectification processes of labor. As Worrell writes, "I must be reduced to a thing, displacing my self for the benefit of the other; to become social or acquire a moral existence, I must will my own dismemberment, my own mortification or self-annihilation."[19] Dead bodies, too, have been displaced, dismembered, and thereby conscripted into laboring for capital. This begs several questions: Does the creation of the funerary gimmick engender the same demand of dead bodies? And if so, what of the remainder—what sorts of identities will dead bodies of the future be able to claim if they refuse to undergo this sublation? What happens to these otherwise abject entities if they fail to be inscribed by the forces of capital? What cremains?

THE BLANK PARODY OF THE BODY

The dead body has become a gimmick. In some places it has even become a celebrity of sorts. It labors much like other celebrities. It is spectacularized and revealed to be just like the rest of us only different; consumed in mass media outlets, it also serves as an object for popular curiosity and exploitation. Bogard, in fact, goes so far as to insist that our society no longer produces corpses so much as it does zombies.[20] And these living dead often make the news as the butt of jokes.

The September 1998 issue of the *Atlantic Monthly* reported that a Kentucky bookbinder had begun selling memorials for the dead. The entrepreneur had found a way to combine paper pulp and cremated human remains to create pages that could then be bound in book volumes. Having apparently discovered a way to capitalize on the growing market for unique funerary goods, he'd labeled his invention "bibliocadavers." The author did not miss an opportunity to quip. "The advent of the bibliocadaver will, if nothing else, add a new facet to the idea of books being remaindered." Though it took a few years, it was finally discovered that the whole bibliocadaver idea was nothing more than a prank the bookbinder was trying to pull on his customers that took on a life of its own. Before the bibliocadaver was revealed as a hoax, the product received little more than a collective shrug from funeral insiders who have grown accustomed to new and wildly inventive products and services appearing on the market every month. The jaded reception by the trade showcases the rapidity of product differentiation in the industry as well as the way novelty for its own sake so quickly wears out its welcome as people move on to the next exciting gimmick.

The *Atlantic Monthly* story represents the parodic impulse (which is accurate, especially given that the bibliocadaver was never actually manufactured). Bookseller Timothy Hawley's action, which is itself a kind of gimmick, calls one's attention to the liminality intrinsic to the gimmick. Like Japanese *koans*, it poses unanswerable questions: Are such parodies funny? Who is the target of the joke (the producer, the consumer, or the media)? And why is the parody necessary? In other words, one could understand parody as revealing a kind of public liminality—a messy and contested terrain where actors compete to define what gets treated as serious, and/or sacred, and what is worthy of satire.

Joey Scaggs is a multimedia artist who also fooled a roster of media outlets (*Los Angeles Times, Boston Herald,* and *Mother Jones* among others) by creating an elaborate website (www.finalcurtain.com) detailing the financing and imminent construction of "a world-wide memorial theme park mall and timeshare operation for artists and creative people" that was set to be franchised around the globe. The hoax was intended to parody the industry and its funerary gimmicks. Given industry actors' apparent eagerness to produce commodity gimmicks, combined with the media's appetite for spectacle, the hoax was evidently easy to pull off. It is not that journalists and their readers are gullible. Rather, the presence of the gimmick, with its deceptive claims on desire, has become nearly taken-for-granted.

Pastiche has become rather unordinary, even expected in today's funerary gimmicks. Where Victorians wore jewelry made from the hair of their dead, we have LifeGem. Colonial American families washed and buried their dead. Today, we have commercialized DIY funerals. Sailors send their dead into the depths of the oceans, but even landlocked Americans have Eternal Reefs and Atlantis. The ongoing drive for the expansion of capital can lead to the co-optation of histories and bodies, by substituting traditions borrowed from the past with commodified gimmicks that transform dead flesh into spectacular memorials.

TIME OUT OF JOINT

The production of the funerary gimmick perpetuates liminality by transforming the dead body into a consumable, often one that is parodic. This levels the status of the body to that of any other product and the semblance of the false promises inherent therein. The industrialist-inventor Charles Kettering, in an article entitled "Keep the Consumer Dissatisfied," wrote, "If everyone were satisfied, no one would buy the new thing because no one would want it... You must accept this reasonable dissatisfaction with what you have and buy the new thing."[21] As a consumable, even the dead body is insufficient for the consumer. She is left with a desire for something else, something more. The consumable is intentionally impermanent.

Rather than serving as an entry point into the life *everlasting* (as symbolized by the ostensible permanence of the burial vault, the ceramic urn, the gasket-sealed casket, or the granite headstone), the gimmick paves the way for the body to assume the shelf-life of any other product; the fleeting celebrity of the spectacle, perhaps, or another type of planned obsolescence. It is rooted in the desire (without its satisfaction) for the consumable in conjunction with the desire (again, without consummation of that desire) to relate to someone who is no longer alive. It disrupts temporality.

The dead body-as-gimmick, with its uncertain shelf-life and its just-out-of-reach wish-fulfillment, suspends closure. With its ever-present trace of a life once-lived, permanent loss is suspended because nothing, including absence, is in fact permanent. This corresponds to Zygmunt Bauman's theorization of "liquid modernity" in which the citizen-consumer understands her life *episodically* (as opposed to serially).[22] What is perceived is a here-and-now that is temporally cut off and autonomous from the past and future. Each moment is discrete and unattached from one's history and there are few if any forthcoming consequences. Life is pure synchrony—a marketer's dream. So, Bauman tells us, where death once signaled a final transformation into a new state of being (or existence), it is becoming increasingly difficult to think of death as anything other than a problem of the body, since humans perceive time as a series of ever-present now's. Death is like debt—just one more bump in the road.

The gimmick reinforces the norm that nothing is ever fully or permanently absent, that everything can be present, everything is possible, and everything can be within one's grasp. Even dead, decaying bodies can be transmuted into something else. Therefore, even though our loved ones are buried, they are not gone—they are dead but still present. Loss is simply not possible.

Since nothing may truly be forgotten, there will always be a trace, a residue. One may note here the parallels between capital's (deceptive) promise to make the consumer whole and the very word *remember*, which means "to make whole"; it is the opposite of *dismember*. Therefore, forgetting is not possible in a "prosthetic culture, which, in revealing the constructedness of presence, implicitly denies the legitimacy of absence."[23] Capital will always leave a trace, even in the absence of a person, a human body, and even if that trace is only a promise, because, as twenty-first-century consumers, we must keep returning to the point of purchase.

Where rituals construct time,[24] gimmicks freeze it. In the past, and to a lesser degree in the present, religious logics contained death with their sacralizing rites of passage. Mourners were comforted knowing that death was little more than a corridor to (eternal) life. Medical logics contained death by consigning the dying to the total institutions of hospitals and nursing homes. With death and dying hidden from view, the living could bracket mortality anxieties and focus instead on the present. Now, the logic of late capitalism with the aid of the gimmick, contains not death, nor the dying, nor the putrefying body, but the *finality* of these things by quite literally inscribing capital on dead bodies and turning these into something that can be invested in, traded, and bought and sold on an open market. Death is just another form of presence, and the dead body is the "living absence"[25] of such a presence—a consumable.

CONCLUSION

It is axiomatic that the "healthy" response to a death is to progress through the stages of grief[26] over an appropriate (but not morbidly extended) period of time. The funeral industry played a significant role beginning mid-twentieth century in defining what the appropriate liminal period entailed and by normalizing "healthy" grieving through its expertise at preparing and processing the dead body for viewings. Ostensibly, by seeing the deceased in peaceful repose, mourners could come to terms with the death, adequately process their loss, and begin to rebuild their lives. Now, the funeral industry's wares are changing. The dead body can be anything the consumer wants, including a source of amusement. The dead body can also be just one more sales ploy to generate the desire for just one more good.

Whether due to its price, its novelty, its conspicuousness, or its production technology, the funerary gimmick is a spectacle for our time. This spectacle, which "is the commodity that has left its material body on earth and risen to a new ethereal presence,"[27] is *in* the world but not *of* it. This spectacular specter represents the promise of proximity without allowing for the full realization or consummation that might ordinarily be associated with other forms of intimacy. It is manufactured to appeal to the pervasive sense that something is not quite right, that something is absent but recoverable, and because it presents the dead body as neither here nor there, it is perfectly liminal.

One problem, however, with inhabiting a permanent liminality may be that while time is not in fact as linear as it is conventionally perceived to be, neither is time static. The past can continue to haunt those of us who are living. Zizek writes, "[W]hy do the dead return? . . . *[B]ecause they were not properly buried,* that is, because something went wrong with their obsequies. The return of the dead is a sign of a disturbance in the symbolic rite, in the process of symbolization; the dead return as collectors of some unpaid symbolic debt."[28] The dead body has been displaced from its resting ground. It has quite literally been dismembered in order that capital may promise a re-membering. It has been conscripted, "called up" from its resting place in the name of the gimmick. It is done so that even the dead may labor, so that none may tarry too long. But, as Zizek's quote suggests, there may be consequences. For, as Robert Harrison states, *"Dasein does not die until its remains are disposed of."*[29] Rather than dispose of dead bodies, however, those laboring bodies are drawn in by capital's centripetal pull. They are recycled, or, perhaps more properly, up-cycled into working alongside living bodies.

For the living, the gimmick provides a ritual of consumption and, with it, the promise that the consumer will find wholeness, so that she may be complete. The gimmick becomes the next gift of shopping for oneself, a retail therapy outing, or even the purchase "to end all" purchases. Thus, consumers' participation in the market perpetuates liminality and suspends temporality, as the search for fulfillment always necessitates a "next" purchase. The gimmick is therefore the commodification of a moment that cannot be surpassed. This lends support to Bauman's claim that "the whole of life [has become] a game of bridge-crossing: all bridges seem by and large alike, all are—comfortably—part of one's daily itinerary, so that no bridge seems to loom ominously as the 'ultimate' one (most importantly, none seems to be the bridge 'of no return')."[30] Every bridge can be recrossed, every sequence is repeatable, and every body can be reanimated to/for the pleasures of consumer desire.

The desire for wholeness that constitutes the subjectivity many of us know may, of course, never be realized. Yet, the gimmick exists as a tangible good that has incorporated an "other" who was formerly both a living and a dead body. But where this commodity implicitly promises wholeness and completion, the very most one can expect is solipsism. Perhaps this is *a propos* in a culture that both fetishizes the body and sacralizes the self. The self-referentiality delivered by the gimmick can accurately be described then as a self-*reverentiality*, wherein the gimmick has become just one more means through which to work on a life project that has no terminus.

NOTES

1. Bruno Gulli, *Labor of Fire: The Ontology of Labor between Economy and Culture* (Philadelphia, PA: Temple University Press, 2005), 2.
2. John S. Baglow, "The Rights of the Corpse," *Mortality* 12 (2007): 224.
3. Victor Turner, "Dewey, Dilthey, and Drama: An Essay in the Anthropology of Experience," in *The Anthropology of Experience*, eds. V. Turner and E. Bruner (Champaign, IL: University of Illinois Press, 1986), 33–44, 419.
4. Cremation Association of North America, *Final 2005 Statistics and Projections to the Year 2025: 2006 Preliminary Data* (Chicago, IL: Market Research and Statistics, 2007).
5. Lisa T. Cullen, *Remember Me: A Lively Tour of the New American Way of Death* (New York: Collins, 2006).
6. Oliver McRae, "Deathcare: Past, Present…and Future," *The Director*, July (2004): 35.

7. Guy Debord, *Society of the Spectacle* Trans. K. Knabb (London, UK: Rebel Press, 1967), 9.
8. Jean Baudrillard, *America,* trans. C. Turner (London, UK: Verso, 1998). *Simulacra and Simulation,* trans. S. Glaser (Ann Arbor: University of Michigan Press [1994] 2006).
9. Zygmunt Bauman, *Mortality, Immortality and Other Life Strategies* (Stanford, CA: Stanford University Press, 1992).
10. David Harvey, "The Body as an Accumulation Strategy," *Economy and Planning* 16 (1998): 406.
11. Robert P. Harrison, *The Dominion of the Dead* (Chicago: University of Chicago Press, 2003), 65.
12. Antonius Robben, "Death and Anthropology: An Introduction," in *Death, Mourning, and Burial: A Cross-Cultural Reader,* ed. A. Robben (Malden, MA: Blackwell, 2004), 1–16, 7–10.
13. James W. Gentry et al., "The Vulnerability of those Grieving the Death of a Loved One: Implications for Public Policy," *Journal of Public Policy and Marketing* 13 (1994): 135.
14. Slavoj Zizek, *Looking Awry: An Introduction to Jacques Lacan through Popular Culture* (Cambridge, MA: MIT Press, 1991), 22.
15. David W. Moller, *Confronting Death: Values, Institutions, and Human Mortality* (New York: Oxford University Press, 1996), 97.
16. Bronna Romanoff and Marion Terenzio, "Rituals and the Grieving Process," *Death Studies* 22 (1998): 699.
17. Baglow, "The Rights of the Corpse," 230.
18. Slavoj Zizek, *Welcome to the Desert of the Real!* (London, UK: Verso, 2002), 10–11.
19. Mark Worrell, *Dialectic of Solidarity: Labor, Antisemitism, and the Frankfurt School* (Leiden, The Netherlands: Brill, 2008), 264.
20. William Bogard, "Empire of the Living Dead," *Mortality* 13 (2008).
21. Charles Kettering, "Keep the Consumer Dissatisfied," *Nation's Business,* 17 (1929): 31.
22. Bauman, *Mortality, Immortality and Other Life Strategies.*
23. Laura E. Tanner, *Lost Bodies: Inhabiting the Borders of Life and Death* (Ithaca, NY: Cornell University Press, 2006), 222.
24. Roy Rappaport, *Ritual and Religion in the Making of Humanity* (Cambridge: Cambridge University Press, 1999).
25. Baglow, "The Rights of the Corpse," 230.
26. See, e.g., Elizabeth Kulber-Ross, *On Death and Dying* (New York: Scribner, 1997).
27. Edward Ball in Mark Worrell, "The Cult of Exchange Value and the Critical Theory of Spectacle," *Fast Capitalism* 5, no.2 (1999).
28. Zizek, *Looking Awry,* 23.
29. Harrison, *The Dominion of the Dead,* 143.
30. Bauman, *Mortality, Immortality and Other Life Strategies,* 173.

INDEX